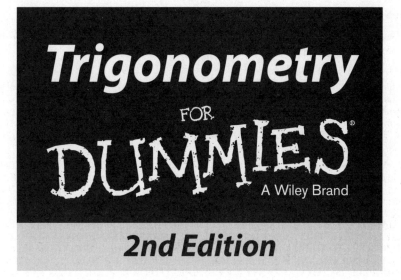

Trigonometry
FOR DUMMIES®
A Wiley Brand

2nd Edition

by Mary Jane Sterling

FOR DUMMIES®
A Wiley Brand

Trigonometry For Dummies,® 2nd Edition

Published by: **John Wiley & Sons, Inc.,** 111 River Street, Hoboken, NJ 07030-5774, www.wiley.com

Copyright © 2014 by John Wiley & Sons, Inc., Hoboken, New Jersey

Published simultaneously in Canada

For general information on our other products and services, please contact our Customer Care Department within the U.S. at 877-762-2974, outside the U.S. at 317-572-3993, or fax 317-572-4002. For technical support, please visit www.wiley.com/techsupport.

Wiley publishes in a variety of print and electronic formats and by print-on-demand. Some material included with standard print versions of this book may not be included in e-books or in print-on-demand. If this book refers to media such as a CD or DVD that is not included in the version you purchased, you may download this material at http://booksupport.wiley.com. For more information about Wiley products, visit www.wiley.com.

Library of Congress Control Number: 2013954199

ISBN 978-1-118-82741-3 (pbk); ISBN 978-1-118-82746-8 (ebk); ISBN 978-1-118-82757-4 (ebk)

Manufactured in the United States of America

10 9 8 7 6 5 4 3 2

Contents at a Glance

Table of Contents

Introduction

• •

*M*any of the more practical and exciting accomplishments of early man were performed using trigonometry. Even before trigonometry was formalized into a particular topic to study or used to solve problems, trigonometry helped people to sail across large bodies of water, build gigantic structures, plot out land, and measure heights and distances — even to the stars.

We still use trigonometry for all these reasons and more. If you're going to get your pilot's license, you'll need trigonometry. Trigonometry is also the basis for many courses in mathematics — starting in grade school with geometric shapes and map reading and moving on through calculus. Trig is all over the place.

You can get as deeply into this topic or as little into it as you want, and you'll still come out of it thinking, "Gee, I didn't realize that trigonometry was used to do this! Wasn't that just loads of fun!" Well, maybe I'm pushing it a bit — *loads* may be a slight exaggeration.

Whether you're pursuing trigonometry so that you can go on in calculus or prepare for architecture or drafting or do that piloting thing, or even if you're just curious, you'll find what you need here. You can get as technical as you want. You can skip through the stuff you don't need. Just know that you'll be on the same adventure as that early man — you'll just have the advantage of a few more tools.

About This Book

So, what's in it for you? What's in a book on trigonometry that'll ring your bell or strike your fancy or just make you pretty happy? Where do I begin?

You can start anywhere in the book, jump around, and just go any direction you please. If you're really excited about triangles and how they can be used to your advantage, check them out. Everything you need to read about them is here.

What if you've got another angle? Or, maybe you didn't have one to begin with but wish you did. If you're looking for angles, you've come to the right place. There are big angles and little angles all named depending on their situation. They're measured in degrees or radians. "What's a radian?" you ask. You can find it in this book, that's for sure.

You may be very analytically minded. If so, you'll find your favorite spot is among all the identities and equation solving. Hop right to it. They're waiting for you.

And if drawing pictures is your bag, go to the chapters on graphing to see what can be done with simple trig graphs, complicated trig graphs, and everything in between. There are even explanations on what the function equations mean, why they're used in an application, and how they're related to the graphs.

You'll find many sidebars throughout this book. Sidebars (gray boxes of text) are those fun little anecdotes that don't really contain a lot of math content but present interesting little tidbits — fun things to read. Neither the sidebars nor the items marked with the Technical Stuff icon are necessary for your understanding of the material. Think of them as little diversions for your reading pleasure.

Within this book, you may note that some web addresses break across two lines of text. If you're reading this book in print and want to visit one of these web pages, simply key in the web address exactly as it's noted in the book, pretending as though the line break doesn't exist. If you're reading this as an e-book, you've got it easy — just click the web address to be taken directly to the web page.

Foolish Assumptions

What kind of fool am I to assume that you're reading _Trigonometry For Dummies_ because it looks more interesting than the latest bestseller? I'm not that foolish! To be honest, trig wouldn't be my first choice for a fun read. I'm just going to assume that you really want to do this. While writing this book, I made a few other assumptions about you as well:

- ✔ You have a goal in mind. You want to conquer some of the topics in this book so you're prepared for a course of study.

- ✔ You have a pretty solid grasp on algebra and can solve a simple algebraic equation without falling completely apart.

- ✔ You're planning on being on a quiz show, and you need to bone up on the possible trig questions.

Icons Used in This Book

Icons are easy to spot. They could be called *eye-cons,* because they catch your eye. Here are the ones I use in this book:

 Of course, trig rules — it's fun! But taken another way, this icon is used to point out to you when particular equations or expressions or formulas are used in trigonometry that you should be paying attention to. They're important. This icon helps you find them again, if you need them.

This icon refers back to information that I think you may already know. It needs to be pointed out or repeated so that the current explanation makes sense.

What about trigonometry isn't technical? Actually, there's quite a bit, but this icon points out the rules or absolutely unchangeable stuff that you may need to understand the situation.

There are always things that are tricky or confusing or problems that just ask for an error to happen. This icon is there to alert you, hoping to help you avoid a mathematical pitfall.

Beyond the Book

In addition to the material in the print or e-book you're reading right now, this product also comes with some access-anywhere goodies on the web. No matter how diligent you are about reading through this material, you'll likely come across a few questions where you don't have a clue. Check out the free Cheat Sheet at www.dummies.com/cheatsheet/trigonometry for helpful information, all provided in a concise, quick-access format.

Also, if you want some practice problems, be sure to find a copy of *Trigonometry Workbook For Dummies.* The problems follow the material in this book and provide some more practice and insights into the processes involved in trigonometry.

Finally, you can find some articles online that tie together and offer new insights to the material you find in this book. Go to www.dummies.com/extras/trigonometry for these informative articles.

Where to Go from Here

Back when I was in college, my friend, Judy Christopher, once consoled me with, "Life is like a sine curve. It has its ups and downs. If you're feeling like you're really down, then just remember that you'll be going up that same amount someday soon." So, if you're in the dumps, maybe you want to start with the graphs of the sine curves and other trig curves. Make of them what you will.

Or maybe, like me, you're a puzzle buff. I can't wait to tackle the Sunday crossword puzzle. You have to call up bits and pieces of information and make them all fit into something logical. If that's what you're interested in today, then go to proving identities and solving equations. That's a great challenge for a rainy, Sunday afternoon's pleasure.

Are you into angles and directions and plans? You may want to start with the ways that angles are measured and how they all fit together in the big picture. The basics are always a good place to start when you're investigating a topic.

No matter where you start with this book, be ready to flip the pages front to back or back to front. Think of it as an adventure that can take you many interesting places. Enjoy!

Part I
Getting Started with Trigonometry

In this part...

- ✔ Become acquainted with angle measures and how they relate to trig functions.

- ✔ Discover formulas that provide lengths of segments, midpoints, and slopes of lines.

- ✔ Become familiar with circles and the relationships between radii, diameters, centers, and arcs.

- ✔ Relate infinitely many angle measures to just one reference angle.

- ✔ Find a simple conversion method for changing from degrees to radians and vice versa.

- ✔ Observe the properties of special right triangles, and use the Pythagorean theorem to formulate the relationships between the sides of these right triangles.

Chapter 1

Troucing Trig Technicalities

*H*ow did Columbus find his way across the Atlantic Ocean? How did the Egyptians build the pyramids? How did early astronomers measure the distance to the moon? No, Columbus didn't follow a yellow brick road. No, the Egyptians didn't have LEGO instructions. And, no, there isn't a tape measure long enough to get to the moon. The common answer to all these questions is trigonometry.

Trigonometry is the study of angles and triangles and the wonderful things about them and that you can do with them. For centuries, humans have been able to measure distances that they can't reach because of the power of this mathematical subject.

Taking Trig for a Ride: What Trig Is

"What's your angle?" That question isn't a come-on such as "What's your astrological sign?" In trigonometry, you measure angles in both degrees and radians. You can shove the angles into triangles and circles and make them do special things. Actually, angles drive trigonometry. Sure, you have to consider algebra and other math to make it all work. But you can't have trigonometry without angles. Put an angle into a trig function, and out pops a special, unique number. What do you do with that number? Read on, because that's what trig is all about.

Sizing up the basic figures

Segments, rays, and lines are some of the basic forms found in geometry, and they're almost as important in trigonometry. As I explain in the following sections, you use those segments, rays, and lines to form angles.

Drawing segments, rays, and lines

A *segment* is a straight figure drawn between two endpoints. You usually name it by its endpoints, which you indicate by capital letters. Sometimes, a single letter names a segment. For example, in a triangle, a lowercase letter may refer to a segment opposite the angle labeled with the corresponding uppercase letter.

A *ray* is another straight figure that has an endpoint on one end, and then it just keeps going forever in some specified direction. You name rays by their endpoint first and then by any other point that lies on the ray.

A *line* is a straight figure that goes forever and ever in either direction. You only need two points to determine a particular line — and only one line can go through both of those points. You can name a line by any two points that lie on it.

Figure 1-1 shows a segment, ray, and line and the different ways you can name them using points.

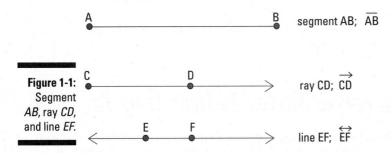

Figure 1-1: Segment *AB*, ray *CD*, and line *EF*.

segment AB; $\overline{AB}$

ray CD; $\overrightarrow{CD}$

line EF; $\overleftrightarrow{EF}$

Intersecting lines

When two lines intersect — if they do intersect — they can only do so at one point. They can't double back and cross one another again. And some curious things happen when two lines intersect. The angles that form between those two lines are related to one another. Any two angles that are next to one another and share a side are called *adjacent angles*. In Figure 1-2, you see several sets of intersecting lines and marked angles. The top two figures indicate two pairs of adjacent angles. Can you spot the other two pairs? The angles that are opposite one another when two lines intersect also have a special name. Mathematicians call these angles *vertical angles*. They don't have a side in common. You can find two pairs of vertical angles in Figure 1-2, the two middle figures indicate the only pairs of vertical angles. Vertical angles are always equal in measure.

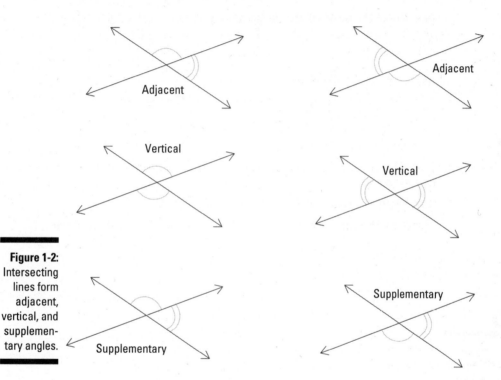

Figure 1-2: Intersecting lines form adjacent, vertical, and supplementary angles.

Why are these different angles so special? They're different because of how they interact with one another. The adjacent angles here are called *supplementary angles*. The sides that they don't share form a straight line, which has a measure of 180 degrees. The bottom two figures show supplementary angles. Note that these are also adjacent.

Angling for position

When two lines, segments, or rays touch or cross one another, they form an angle or angles. In the case of two intersecting lines, the result is four different angles. When two segments intersect, they can form one, two, or four angles; the same goes for two rays.

These examples are just some of the ways that you can form angles. Geometry, for example, describes an angle as being created when two rays have a common endpoint. In practical terms, you can form an angle in many ways, from many figures. The business with the two rays means that you can extend the two sides of an angle out farther to help with measurements, calculations, and practical problems.

Describing the parts of an angle is pretty standard. The place where the lines, segments, or rays cross is called the *vertex* of the angle. From the vertex, two sides extend.

Naming angles by size

You can name or categorize angles based on their size or measurement in degrees (see Figure 1-3):

- **Acute:** An angle with a positive measure less than 90 degrees
- **Obtuse:** An angle measuring more than 90 degrees but less than 180 degrees
- **Right:** An angle measuring exactly 90 degrees
- **Straight:** An angle measuring exactly 180 degrees (a straight line)
- **Oblique:** An angle measuring more than 180 degrees

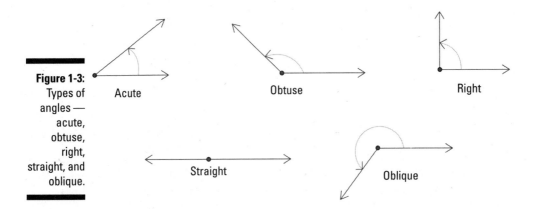

Figure 1-3:
Types of angles —
acute, obtuse, right, straight, and oblique.

Acute

Obtuse

Right

Straight

Oblique

Naming angles by letters

How do you name an angle? Why does it even need a name? In most cases, you want to be able to distinguish a particular angle from all the others in a picture. When you look at a photo in a newspaper, you want to know the names of the different people and be able to point them out. With angles, you should feel the same way.

You can name an angle in one of three different ways:

- ✔ **By its vertex alone:** Often, you name an angle by its vertex alone because such a label is efficient, neat, and easy to read. In Figure 1-4, you can call the angle *A*.

- ✔ **By a point on one side, followed by the vertex, and then a point on the other side:** For example, you can call the angle in Figure 1-4 angle *BAC* or angle *CAB*. This naming method is helpful if someone may be confused as to which angle you're referring to in a picture. ***Remember:*** Make sure you always name the vertex in the middle.

- ✔ **By a letter or number written inside the angle:** Usually, that letter is Greek; in Figure 1-4, however, the angle has the letter *w*. Often, you use a number for simplicity if you're not into Greek letters or if you're going to compare different angles later.

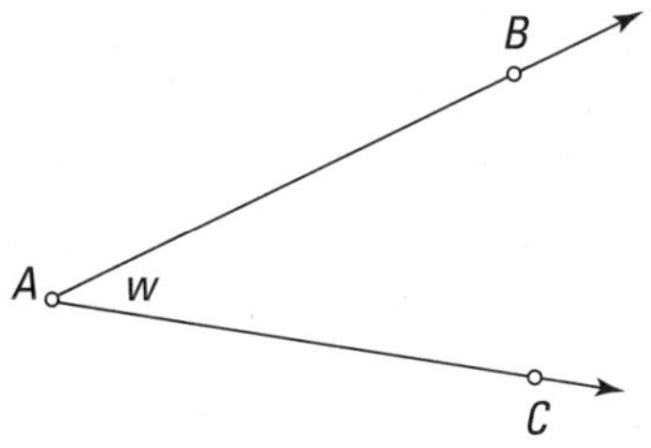

Figure 1-4:
Naming an
angle.

Triangulating your position

All on their own, angles are certainly very exciting. But put them into a triangle, and you've got icing on the cake. Triangles are one of the most frequently studied geometric figures. The angles that make up the triangle give them many of their characteristics.

Angles in triangles

A triangle always has three angles. The angles in a triangle have measures that always add up to 180 degrees — no more, no less. A triangle named *ABC* has angles *A*, *B*, and *C*, and you can name the sides $\overline{AB}$, $\overline{BC}$, and $\overline{AC}$, depending on which two angles the side is between. The angles themselves can be acute, obtuse, or right. If the triangle has either an obtuse or right angle, then the other two angles have to be acute.

Naming triangles by their shape

Triangles can have special names based on their angles and sides. They can also have more than one name — a triangle can be both acute and isosceles, for example. Here are their descriptions, and check out Figure 1-5 for the pictures:

- ✔ **Acute triangle:** A triangle where all three angles are acute.

- ✔ **Right triangle:** A triangle with a right angle (the other two angles must be acute).

- ✔ **Obtuse triangle:** A triangle with an obtuse angle (the other two angles must be acute).

- ✔ **Isosceles triangle:** A triangle with two equal sides; the angles opposite those sides are equal, too.

- ✔ **Equilateral triangle:** A triangle where all three side lengths are equal; all the angles measure 60 degrees, too.

- ✔ **Scalene triangle:** A triangle with no angles or sides of the same measure.

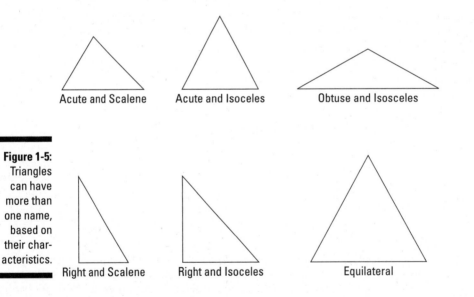

Figure 1-5:
Triangles can have more than one name, based on their characteristics.

Acute and Scalene	Acute and Isoceles	Obtuse and Isoceles
Right and Scalene	Right and Isoceles	Equilateral

Circling the wagons

A *circle* is a geometric figure that needs only two parts to identify it and classify it: its *center* (or middle) and its *radius* (the distance from the center to any point on the circle). Technically, the center isn't a part of the circle; it's just a sort of anchor or reference point. The circle consists only of all those points that are the same distance from the center.

Radius, diameter, circumference, and area

After you've chosen a point to be the center of a circle and know how far that point is from all the points that lie on the circle, you can draw a fairly decent picture. With the measure of the radius, you can tell a lot about the circle: its *diameter* (the distance from one side to the other, passing through the center), its *circumference* (how far around it is), and its *area* (how many square inches, feet, yards, meters — what have you — fit into it). Figure 1-6 shows these features.

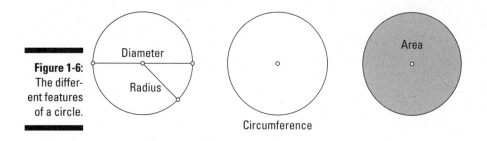

Figure 1-6: The different features of a circle.

Ancient mathematicians figured out that the circumference of a circle is always a little more than three times the diameter of a circle. Since then, they narrowed that "little more than three times" to a value called *pi* (pronounced "pie"), designated by the Greek letter π. The decimal value of π isn't exact — it goes on forever and ever, but most of the time, people refer to it as being approximately 3.14 or $\frac{22}{7}$, whichever form works best in specific computations.

The formula for figuring out the circumference of a circle is tied to π and the diameter:

Circumference of a circle: $C = \pi d = 2\pi r$

The d represents the measure of the diameter, and r represents the measure of the radius. The diameter is always twice the radius, so either form of the equation works.

Similarly, the formula for the area of a circle is tied to π and the radius:

Area of a circle: $A = \pi r^2$

This formula reads, "Area equals pi are squared." And all this time, I thought that *pies are round.*

Don't give me that *jiva*

The ancient Greek mathematician Ptolemy was born some time at the end of the first century. Ptolemy based his version of trigonometry on the relationships between the chords of circles and the corresponding central angles of those chords. Ptolemy came up with a theorem involving four-sided figures that you can construct with the chords. In the meantime, mathematicians in India decided to use the measure of *half* a chord and *half* the angle to try to figure out these relationships. Drawing a radius from the center of a circle through the middle of a chord (halving it) forms a right angle, which is important in the definitions of the trig functions. These half-measures were the beginning of the sine function in trigonometry. In fact, the word *sine* actually comes from the Hindu name *jiva*.

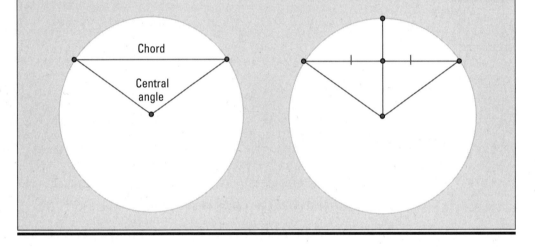

Example: Find the radius, circumference, and area of a circle if its diameter is equal to 10 feet in length.

If the diameter (d) is equal to 10, you write this value as $d = 10$. The radius is half the diameter, so the radius is 5 feet, or $r = 5$. You can find the circumference by using the formula $C = \pi d = \pi \cdot 10 \approx 3.14 \cdot 10 = 31.4$. So, the circumference is about $31\frac{1}{2}$ feet around. You find the area by using the formula $A = \pi r^2 = \pi \cdot 5^2 = \pi \cdot 25 \approx 3.14 \cdot 25 \approx 78.5$, so the area is about $78\frac{1}{2}$ square feet.

Chord versus tangent

You show the diameter and radius of a circle by drawing segments from a point on the circle either to or through the center of the circle. But two other straight figures have a place on a circle. One of these figures is called a chord, and the other is a tangent:

✓ **Chords:** A *chord* of a circle is a segment that you draw from one point on the circle to another point on the circle (see Figure 1-7). A chord always stays inside the circle. The largest chord possible is the diameter — you can't get any longer than that segment.

✓ **Tangent:** A *tangent* to a circle is a line, ray, or segment that touches the outside of the circle in exactly one point, as in Figure 1-7. It never crosses into the circle. A tangent can't be a chord, because a chord touches a circle in two points, crossing through the inside of the circle. Any radius drawn to a tangent is perpendicular to that tangent.

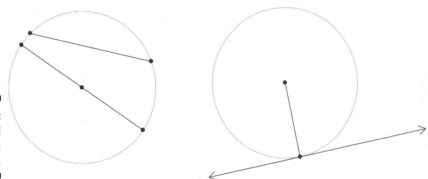

Figure 1-7:
Chords and
tangent of a
circle.

Angles in a circle

There are several ways of drawing an angle in a circle, and each has a special way of computing the size of that angle. Four different types of angles are: central, inscribed, interior, and exterior. In Figure 1-8, you see examples of these different types of angles.

Central angle

A *central angle* has its vertex at the center of the circle, and the sides of the angle lie on two radii of the circle. The measure of the central angle is the same as the measure of the arc that the two sides cut out of the circle.

Inscribed angle

An *inscribed angle* has its vertex on the circle, and the sides of the angle lie on two chords of the circle. The measure of the inscribed angle is half that of the arc that the two sides cut out of the circle.

Interior angle

An *interior angle* has its vertex at the intersection of two lines that intersect inside a circle. The sides of the angle lie on the intersecting lines. The measure of an interior angle is the average of the measures of the two arcs that are cut out of the circle by those intersecting lines.

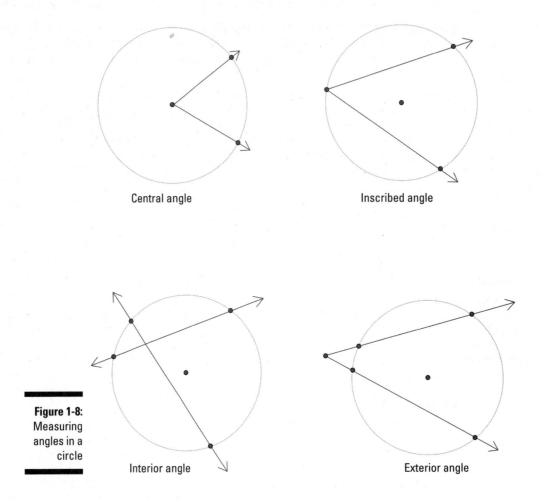

Central angle

Inscribed angle

Figure 1-8:
Measuring
angles in a
circle

Interior angle

Exterior angle

Exterior angle

An *exterior angle* has its vertex where two rays share an endpoint outside a circle. The sides of the angle are those two rays. The measure of an exterior angle is found by dividing the difference between the measures of the intercepted arcs by two.

Example: Find the measure of angle *EXT*, given that the exterior angle cuts off arcs of 20 degrees and 108 degrees (see Figure 1-9).

Find the difference between the measures of the two intercepted arcs and divide by 2:

$$\frac{108-20}{2} = \frac{88}{2} = 44$$

The measure of angle *EXT* is 44 degrees.

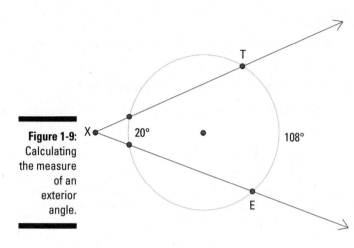

Figure 1-9:
Calculating
the measure
of an
exterior
angle.

Sectioning sectors

A *sector* of a circle is a section of the circle between two *radii* (plural for radius). You can consider this part like a piece of pie cut from a circular pie plate (see Figure 1-10).

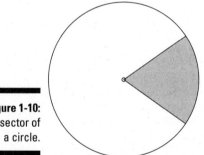

Figure 1-10:
A sector of
a circle.

You can find the area of a sector of a circle if you know the angle between the two radii. A circle has a total of 360 degrees all the way around the center, so if that central angle determining a sector has an angle measure of 60 degrees, then the sector takes up $\frac{60}{360}$, or $\frac{1}{6}$, of the degrees all the way around. In that case, the sector has $\frac{1}{6}$ the area of the whole circle.

Example: Find the area of a sector of a circle if the angle between the two radii forming the sector is 80 degrees and the diameter of the circle is 9 inches.

1. **Find the area of the circle.**

   The area of the whole circle is $A = \pi r^2 = \pi \cdot (4.5)^2 \approx 3.14(20.25) \approx 63.585$, or about $63\frac{1}{2}$ square inches.

2. **Find the portion of the circle that the sector represents.**

   The sector takes up only 80 degrees of the circle. Divide 80 by 360 to get

   $$\frac{80}{360} = \frac{2}{9} \approx 0.222.$$

3. **Calculate the area of the sector.**

   Multiply the fraction or decimal from Step 2 by the total area to get the area of the sector: $0.222(63.585) \approx 14.116$. The whole circle has an area of almost 64 square inches, and the sector has an area of just over 14 square inches.

Understanding Trig Speak

Any math or science topic has its own unique vocabulary. Some very nice everyday words have new and special meanings when used in the context of that subject. Trigonometry is no exception.

Defining trig functions

Every triangle has six parts: three sides and three angles. If you measure the sides and then pair up those measurements (taking two at a time), you have three different pairings. Do division problems with the pairings — changing the order in each pair — and you have six different answers. These six different answers represent the six trig functions. For example, if your triangle has sides measuring 3, 4, and 5, then the six divisions are $\frac{3}{4}, \frac{4}{3}, \frac{3}{5}, \frac{5}{3}, \frac{4}{5}$, and $\frac{5}{4}$.

In Chapter 7, you find out how all these fractions work in the world of trig functions by using the different sides of a right triangle. And then, in Chapter 8, you take a whole different approach as you discover how to define the trig functions with a circle.

The six trig functions are named *sine, cosine, tangent, cotangent, secant,* and *cosecant.* Many people confuse the spoken word *sine* with *sign* — you can't really tell the difference when you hear it unless you're careful with the context. You can "go off on a tangent" in some personal dealings, but that phrase has a whole different meaning in trig. Cosigning a loan isn't what trig has in mind, either. The other three ratios are special to trig speak — you can't confuse them with anything else.

Interpreting trig abbreviations

Even though the word *sine* isn't all that long, you have a three-letter abbreviation for this trig function and all the others. Mathematicians find using abbreviations easier, and those versions fit better on calculator keys. The functions and their abbreviations are

sine: sin	**cosine:** cos
tangent: tan	**cotangent:** cot
secant: sec	**cosecant:** csc

As you can see, the first three letters in the full name make up the abbreviations, except for cosecant's.

Noting notation

Angles are the main focus in trigonometry, and you can work with them even if you don't know their measure. Many angles and their angle measures have general rules that apply to them. You can name angles by one letter, three letters, or a number, but to do trig problems and computations, mathematicians commonly refer to the angle names and their measures with Greek letters.

The most commonly used letters for angle measures are α (alpha), β (beta), γ (gamma), and θ (theta). Also, many equations use the variable x to represent an angle measure.

Algebra has conventional notation involving superscripts, such as the 2 in x^2. In trigonometry, superscripts have the same rules and characteristics as in other mathematics. But trig superscripts often look very different. Table 1-1 presents a listing of many of the ways that trig uses superscripts.

Table 1-1	How You Use Superscripts in Trig	
How to Write in Trig Notation	*Alternate Notation*	*What the Superscript Means*
$\sin^2 \theta$	$(\sin \theta)^2$	Square the sine of the angle theta
$(\sin \theta)^{-1}$	$\dfrac{1}{\sin \theta}$	Find the reciprocal of the sine of theta
$\sin^{-1} \theta$	$\arcsin \theta$	Find the angle theta given its sine

The first entry in Table 1-1 shows how you can save having to write parentheses every time you want to raise a trig function to a power. This notation is neat and efficient, but it can be confusing if you don't know the "code." The second entry shows you how to write the reciprocal of a trig function. It means you should take the value of the function and divide it into the number 1. The last entry in Table 1-1 shows how you write the *inverse sine* function. Using the –1 superscript between sine and the angle means that you're talking about inverse sine (or *arcsin*), not the reciprocal of the function. In Chapter 15, I cover the inverse trig functions in great detail, making this business about the notation for an inverse trig function even more clear.

Functioning with angles

The functions in algebra use many operations and symbols that are different from the common add, subtract, multiply, and divide signs in arithmetic. For example, take a look at the square-root operation, $\sqrt{25} = 5$. Putting 25 under the *radical* (square-root symbol) produces an answer of 5. Other operations in algebra, such as absolute value, factorial, and step-function, are used in trigonometry, too. But the world of trig expands the horizon, introducing even more exciting processes. When working with trig functions, you have a whole new set of values to learn or find. For instance, putting 25 into the sine function looks like this: sin 25. The answer that pops out is either 0.423 or –0.132, depending on whether you're using degrees or radians (for more on those two important trig concepts, head on over to Chapters 4 and 5). You can't usually determine or memorize all the values that you get by putting angle measures into trig functions. So, you need trig tables of values or scientific calculators to study trigonometry.

In general, when you apply a trig function to an angle measure, you get some real number (if that angle is in its domain). Some angles and trig functions have nice values, but most don't. Table 1-2 shows the trig functions for a 30-degree angle.

Table 1-2	The Trig Functions for a 30-Degree Angle	
Trig Function	*Exact Value*	*Value Rounded to Three Decimal Places*
sin 30°	$\dfrac{1}{2}$	0.500
cos 30°	$\dfrac{\sqrt{3}}{2}$	0.866
tan 30°	$\dfrac{\sqrt{3}}{3}$	0.577

Trig Function	Exact Value	Value Rounded to Three Decimal Places
cot 30°	$\sqrt{3}$	1.732
sec 30°	$\dfrac{2\sqrt{3}}{3}$	1.155
csc 30°	2	2.000

Some characteristics that the entries in Table 1-2 confirm are that the sine and cosine functions always have values that are between and including –1 and 1. Also, the secant and cosecant functions always have values that are equal to or greater than 1 or equal to or less than –1. (I discuss these properties in more detail in Chapter 7.)

Using the table in the Appendix, you can find more values of trig functions for particular angle measures (in degrees):

$$\tan 45° = 1$$

$$\csc 90° = 1$$

$$\sec 60° = 2$$

I chose these sample values so the answers look nice and whole. Most angles and most functions look much messier than these examples.

Taming the radicals

A *radical* is a mathematical symbol that means, "Find the number that multiplies itself by itself one or more times to give you the number under the radical." You can see why you use a symbol such as $\sqrt{}$ rather than all those words. Radicals represent values of functions that are used a lot in trigonometry. In Chapter 7, I define the trig functions by using a right triangle. To solve for the lengths of a right triangle's sides by using the Pythagorean theorem, you have to compute some square roots, which use radicals. Some basic answers to radical expressions are $\sqrt{16} = 4$, $\sqrt{121} = 11$, $\sqrt[3]{8} = 2$, and $\sqrt[4]{81} = 3$.

These examples are all *perfect squares, perfect cubes,* or *perfect fourth roots,* which means that the answer is a number that ends — the decimal doesn't go on forever. The following section discusses a way to simplify radicals that aren't perfect roots.

Simplifying radical forms

Simplifying a radical form means to rewrite it with a smaller number under the radical — if possible. You can simplify this form only if the number under the radical has a perfect square or perfect cube (or perfect whatever factor) that you can factor out.

Example: Simplify $\sqrt{80}$.

The number 80 isn't a perfect square, but one of its factors, 16, is. You can write the number 80 as the product of 16 and 5, write the two radicals separately, and then evaluate each radical. The resulting product is the simplified form:

$$\sqrt{80} = \sqrt{16 \cdot 5} = \sqrt{16}\sqrt{5} = 4\sqrt{5}$$

Example: Simplify $\sqrt[3]{250}$.

The number 250 isn't a perfect cube, but one of its factors, 125, is. Write 250 as the product of 125 and 2; separate, evaluate, and write the simplified product: $\sqrt[3]{250} = \sqrt[3]{125 \cdot 2} = \sqrt[3]{125}\sqrt[3]{2} = 5\sqrt[3]{2}$.

Approximating answers

As wonderful as a simplified radical is, and as useful as it is when you're doing further computations in math, sometimes you just need to know about how much the value's worth.

Approximating an answer means to shorten the actual value in terms of the number of decimal places. You may find approximating especially useful when the decimal value of a number goes on forever without ending or repeating. When you approximate an answer, you *round* it to a certain number of decimal places, letting the rest of the decimal values drop off. Before doing that, though, you need to consider how big a value you're dropping off. If the numbers that you're dropping off start with a 5 or greater, then bump up the last digit that you leave on by 1. If what you're dropping off begins with a 4 or less, then just leave the last remaining digit alone.

They called this simpler?

Some ancient mathematicians didn't like to write fractions unless they had a numerator of 1. They only liked the fractions $\frac{1}{2}, \frac{1}{3}, \frac{1}{4}, \frac{1}{5}$, and so on. So what did they do when they needed to write the fraction $\frac{5}{6}$? They wrote $\frac{1}{2} + \frac{1}{3}$ instead (because $\frac{1}{2} + \frac{1}{3} = \frac{5}{6}$). What a pain to have to write $\frac{1}{2} + \frac{1}{4} + \frac{1}{10}$ rather than $\frac{17}{20}$. Or maybe you prefer this approach, too?

Example: Round the number 3.141592654 to four decimal places, three decimal places, and two decimal places.

- ✔ **Four decimal places:** This rounding value means that the 3.141? stays (the question mark holds that last place until you make a decision). Because you get to drop off the 92654, and 9 is the first digit of those dropped numbers, bump up the last digit (the 5) to 6. Rounded to four places, 3.141592654 rounds to 3.1416.

- ✔ **Three decimal places:** The 3.14? stays. Because you drop off the 592654, and 5 is the first digit of those numbers, bump up the last digit that you're keeping (the 1) to 2. Rounded to three places, 3.141592654 rounds to 3.142.

- ✔ **Two decimal places:** The 3.1? stays. Because the 1592654 drops off, and 1 is the first digit of those numbers, then the last digit that you're keeping (the 4) stays the same. Rounded to two places, 3.141592654 rounds to 3.14.

Use this technique when approximating radical values. Using a calculator, the decimal value of $\sqrt{80}$ is about 8.94427191. Depending on what you're using this value for, you may want to round it to two, three, four, or more decimal places. Rounded to three decimal places, this number is 8.944.

Equating and Identifying

Trigonometry has the answers to so many questions in engineering, navigation, and medicine. The ancient astronomers, engineers, farmers, and sailors didn't have the current systems of symbolic algebra and trigonometry to solve their problems, but they did well and set the scene for later mathematical developments. People today benefit big-time by having ways to solve equations in trigonometry that are quick and efficient; trig now includes special techniques and identities to fool around — all thanks to the mathematicians of old who created the systems that we use today.

The methods that you use for solving equations in algebra take a completely different turn from usual solutions when you use trig *identities* (in short, equivalences that you can substitute into equations in order to simplify them). To make matters easier (or, some say, to *complicate* them), the different trig functions can be written many different ways. They almost have split personalities. When you're solving trig equations and trig identities, you're sort of like a detective working your way through to substitute, simplify, and solve. What answers should you expect when solving the equations? Why, angles, of course!

Take, for example, one trigonometric equation: $\sin\theta + \cos^2\theta = 1$.

The point of the problem is to figure out what angle or angles should replace the θ to make the equation true. In this case, if θ is 0 degrees, 90 degrees, or 180 degrees, the equation is true.

If you replace θ by 0 degrees in the equation, you get

$$\sin 0° + (\cos 0°)^2 = 1$$
$$0 + (1)^2 = 1$$
$$1 = 1$$

If you replace θ by 90 degrees in the equation, you get

$$\sin 90° + (\cos 90°)^2 = 1$$
$$1 + (0)^2 = 1$$
$$1 = 1$$

Something similar happens with 180 degrees and all the other angle measures that work in this equation (there are an infinite number of solutions). But remember that not just any angle will work here. I carefully chose the angles that are *solutions,* which are the angles that make the equation true. In order to solve trig equations like this one, you have to use inverse trig functions, trig identities, and various algebraic techniques. You can find all the details on how to use these processes in Chapters 11 through 16. And when you've got those parts figured out, dive into Chapter 17, where the equation-solving comes in.

In this particular case, you need to use an *identity* to solve the equation for all its answers. You replace the $\cos^2\theta$ by $1 - \sin^2\theta$ so that all the terms have a sine in them — or just a number. You actually have several other choices for changing the identity of $\cos^2\theta$. I chose $1 - \sin^2\theta$, but some other choices include $\dfrac{1}{\sec^2\theta}$ and $\dfrac{1+\cos 2\theta}{2}$. Discover how to actually solve equations like this one in Chapter 17.

This example just shows you that the identity of the trig functions can change an expression significantly — but according to some very strict rules.

Graphing for Gold

The trig functions have distinctive graphs that you can use to help understand their values over certain intervals and in particular applications. In this section, I describe the axes and show you six basic graphs.

Describing graphing scales

You use the *coordinate plane* for graphing in algebra, geometry, and other mathematical topics. The *x*-axis goes left and right, and the *y*-axis goes up and down. You can also use the coordinate plane in trigonometry, with a little added twist.

The *x*-axis in a trig sketch has tick marks that can represent both numbers (either positive or negative) and angle measures (either in degrees or radians). You usually want the horizontal and vertical tick marks to have the same distance between them. To make equivalent marks on the *x*-axis in degrees, figure that every 90 degrees is about 1.6 units (the same units that you're using on the vertical axis). These *units* represent numbers in the real-number system. This conversion method works because of the relationship between degree measure and radian measure. Check out the method used to do the computation for this conversion in Chapter 5.

Recognizing basic graphs

The graphs of the trig functions have many similarities and many differences. The graphs of the sine and cosine look very much alike, as do the tangent and cotangent, and then the secant and cosecant have similiarities. But those three groupings do look different from one another. The one characteristic that ties them all together is the fact that they're *periodic*, meaning they repeat the same curve or pattern over and over again, in either direction along the *x*-axis. Check out Figures 1-11 through 1-16 to see for yourself.

I say a lot more in this book about the trig-function graphs, and you can find that discussion in Chapters 19, 20, 21, and 22.

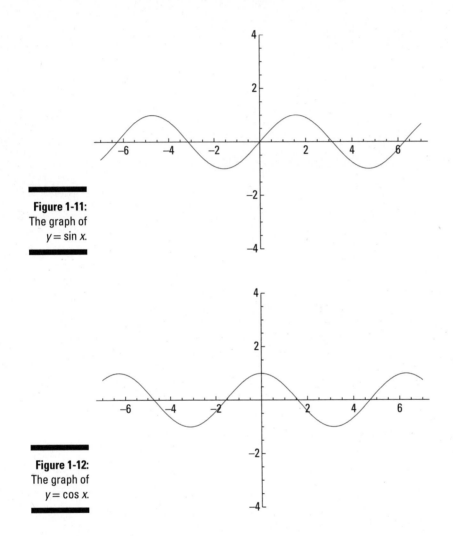

Figure 1-11:
The graph of
$y = \sin x$.

Figure 1-12:
The graph of
$y = \cos x$.

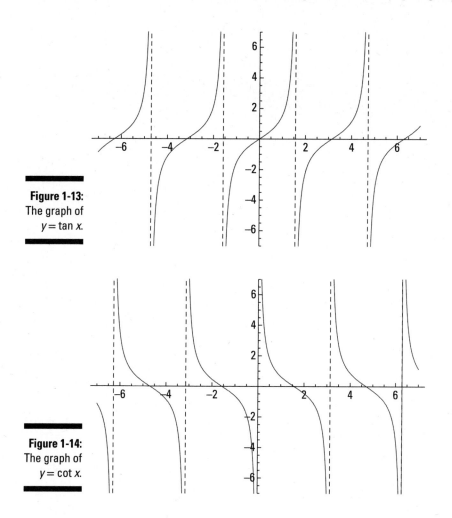

Figure 1-13:
The graph of
$y = \tan x$.

Figure 1-14:
The graph of
$y = \cot x$.

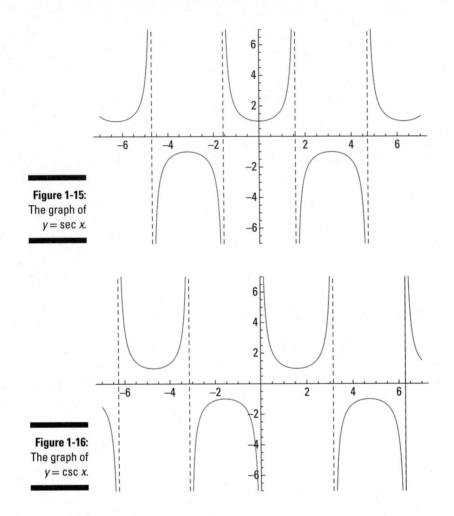

Figure 1-15:
The graph of
$y = \sec x$.

Figure 1-16:
The graph of
$y = \csc x$.

Chapter 2

Coordinating Your Efforts with Cartesian Coordinates

A picture is worth a thousand words. Drawing pictures or graphs of functions and equations in math helps you understand what's going on with them. In trigonometry, you often draw angles and triangles, in addition to the curves that represent the trig functions (sine, cosine, tangent, cotangent, secant, and cosecant). The standard Cartesian coordinate system, which you use when drawing graphs in algebra and other math topics, works best here. If you're looking for a refresher on this point-plotting system, you can find it in this chapter. In short, with the Cartesian coordinate system, everything reads from left to right and from bottom to top, running through the negative to the positive numbers.

Starting Out Simple: Plotting Points

Plotting points on a mathematical graph means finding the correct position for a dot that represents an ordered pair of numbers, such as (2,3), (–1,4), or (0,0). This ordered pair (x,y) is called the *Cartesian coordinates* of the point. You start with two intersecting lines called *axes*.

Axes, axes, we all fall down

Plotting points and drawing graphs requires two axes and a defined distance or scale on them. The two intersecting, perpendicular lines that make up a graph are called the horizontal and vertical *axes* (or *coordinate axes*).

These lines extend left and right, up and down, without end. The horizontal axis is traditionally known as the *x*-axis, although in trigonometry the horizontal axis is often labeled the *θ* axis. The vertical axis is the *y*-axis. The two axes intersect at the *origin,* labeled *O*. The part of the *x*-axis going to the right represents positive numerical values, and you use it as the starting place or initial side when drawing angles in the *standard position.* An angle in standard position has its vertex at the origin, its initial side along the positive *x*-axis, and its terminal side a ray rotated in a counterclockwise direction for positive measures.

Determining the origin of it all

The point where the two axes cross is called the *origin.* You label it with an *O* or with its ordered pair (0,0). The origin is the starting point for counting off the coordinates when graphing all other points. It's also the endpoint of the *rays* (lines that extend infinitely in one direction) that you use when drawing angles in the standard position on the coordinate axes.

Plotting x versus y

Plotting points in a coordinate system involves counting distances to the right or left and up or down from the origin. The axes serve as a starting place. The points are represented or named by the ordered pair of numbers, (*x*,*y*), called the *x*-coordinate and the *y*-coordinate. The designation *ordered pair* means that the order *does* matter. The *x*-coordinate always comes first, and the *y*-coordinate comes last so that this whole graphing system is universal.

Putting da cart before da horse

Rene Descartes was considered to be a mover and shaker in the 17th-century scientific community. He was responsible for many innovations in algebra and geometry. He's also credited with creating our coordinate system used for graphing mathematical objects. The *x*- and *y*-coordinates (the values that specify a location on a graph) are called *Cartesian coordinates* in honor of Descartes.

The *x*-coordinate is the distance to the left or right of the origin that the point lies. If the *x*-coordinate is positive, you move to the right of the origin. If it's negative, you move to the left. The second number, the *y*-coordinate, is the distance up or down from the origin. Positive numbers mean the point is up, and negatives mean you move south of the *x*-axis.

The point (2,4) is two units to the right and four units up from the origin; (–3,2) is three units to the left and two units up; (–4,–3) is four units to the left and three units down; and (5,–1) is five units to the right and one unit down. Points can lie on one of the axes, too. Those points always have a 0 for the *x*- or *y*-coordinate. The point (0,3) lies on the *y*-axis, and (1,0) lies on the *x*-axis. See how to graph all these points in Figure 2-1.

Figure 2-1:
Six points, graphed and labeled.

Cutting the graph into four parts

The intersecting *x*- and *y*-axes divide the whole picture, or *coordinate plane*, into four separate regions called *quadrants*. The quadrants are numbered, starting in the upper-right quadrant and going counterclockwise, as shown in Figure 2-2. Traditionally, you number the quadrants with Roman numerals.

These quadrant number designations are useful when referring to certain types of angles, groupings of points, and trig function properties. The points in Quadrant I all have both *x*- and *y*-coordinates that are positive. In Quadrant II, the *x*-coordinate is negative, and the *y*-coordinate is positive. The points in Quadrant III have both *x*- and *y*-coordinates that are negative. In Quadrant IV, the *x*-coordinate is positive, and the *y*-coordinate is negative.

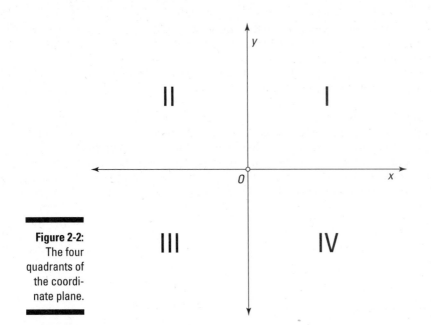

Figure 2-2:
The four
quadrants of
the coordi-
nate plane.

From Here to There: Calculating Distances

The lengths of segments and distances between points play a major part in establishing the trig functions, relationships, and identities (which I cover in Chapters 3 and 11). You can compute these lengths and distances fairly easily, because the coordinate system is just so darned convenient.

Counting on vertical and horizontal distances

When the distance that you're measuring is either vertical or horizontal, then the computation is a simple subtraction problem. One coordinate in each ordered pair is the same. Just find the difference between the other two coordinates.

For instance, to find the distance between the points (5,2) and (5,6), subtract 2 from 6 to get the distance of 4 units between them. This distance is vertical, because the two points have the same x-coordinate, and the second point is directly above the first. To find the distance between the two points (5,6) and (5,–3), subtract –3 from 6 to get a distance of 9 units. Always subtract the smaller number from the larger number, so that the distance you get is a positive number. (Negative distances don't make sense; after all, you can't travel –5 miles to Aunt Myrtle's house!)

TIP

Another way to deal with the different signs of the answers that occur is to use absolute value — then it doesn't matter in what order you subtract the numbers. Take, for example, the preceding example, where I subtracted –3 from 6. The –3 is smaller, so subtracting in that order gave a positive answer. The alternative is to subtract in the opposite order and take the absolute value of the result. If you do –3 – 6, you get –9. The absolute value of –9, written |–9|, equals 9.

Horizontal distances work the same way. In Figure 2-3, you can see the horizontal distance between two points. To find the distance between the points (–8,2) and (5,2), just calculate the difference between –8 and 5, because the *y*-coordinates are the same. The smaller number is –8, so subtracting 5 – (–8), the answer is 13 units. Subtracting in the other order and using absolute value, you get –8 – 5 = –13, and |–13| = 13. You can also see the vertical distance between two points, (5,6) and (5,2), in Figure 2-3. This problem uses simple arithmetic. The difference between 6 and 2 is 4.

Figure 2-3:
Vertical and
horizontal
distances
between
points.

(5,6)

4 units

13 units

(5,2)

(–8,2)

y

x

TRIG RULES

$$\frac{\begin{array}{r}1\\+1\end{array}}{2}$$

Finding the distance between pairs of vertical or horizontal points, (x_1,y_1) and (x_2,y_2), is easy:

✔ Vertical distance (the *x*-coordinates are the same) is $|y_1 - y_2|$.

✔ Horizontal distance (the *y*-coordinates are the same) is $|x_1 - x_2|$.

Another slant: Diagonal distances

Sometimes the distances or lengths you want to determine are on a slant — they go diagonally from one point to another. The formula for determining these distances is based on the Pythagorean theorem.

The Pythagorean theorem

Way back when, Pythagoras discovered a relationship between the sides of any *right triangle,* where one angle is 90 degrees, as Figure 2-4 shows.

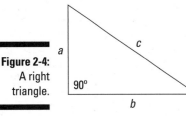

Figure 2-4:
A right
triangle.

Pythagoras found that if *a* and *b* are the lengths of the shorter sides of the right triangle, and if *c* is the length of the *hypotenuse* (the side opposite the right angle), then $a^2 + b^2 = c^2$. You can use this formula to find the diagonal distances between two points on a graph, because the horizontal and vertical distances, which are the sides of the triangle, are easy to find in a coordinate system.

Determining diagonal distances

Using the Pythagorean theorem, solve for *c,* the length of the hypotenuse of a right triangle, and you get

$$c = \sqrt{a^2 + b^2}$$

If the length *a* is the horizontal distance, then you calculate that distance by subtracting the *x*-coordinates; if length *b* is the vertical distance, you get it by subtracting the *y*-coordinates.

To get the general distance formula, simply substitute the difference between the *x*- and *y*-values for *a* and *b* in the Pythagorean theorem, and use the variable *d* (meaning distance) in place of *c.*

The distance, d, between two points (x_1, y_1) and (x_2, y_2) is

$$d = \sqrt{(x_1 - x_2)^2 + (y_1 - y_2)^2}$$

For example, follow these steps to find the distance between the points (3,–4) and (–2,5):

1. **Replace x_1 and x_2 with 3 and –2. Replace the y_1 and y_2 with –4 and 5.**

$$d = \sqrt{(3 - (-2))^2 + (-4 - 5)^2}$$

2. **Subtract the coordinates.**

$$= \sqrt{5^2 + (-9)^2}$$

3. **Add the results and find the square root, if possible.**

$$= \sqrt{25 + 81} = \sqrt{106}$$

In the preceding example, the number under the radical isn't a perfect square. You can either leave the answer with the square-root symbol or give a decimal approximation (see the following section). To three decimal places, the distance in this example is 10.296 units.

When you're calculating the distance between two points, it doesn't matter in what order you subtract the points, as long as you subtract x from x and y from y. Squaring the differences will always result in a positive answer, anyway.

Using exact values or estimating distances

Calculating the distance between two points often leaves you with the square root of a number that isn't a perfect square; this type of answer is called an *irrational number.* Writing the number with the square-root symbol, for example, $\sqrt{47}$, is considered to be writing the *exact value* of the distance. Using a calculator to find a decimal approximation doesn't give an exact answer, because the decimal values go on forever and ever and never repeat in a pattern. Because the decimals are always estimates, mathematicians often insist on leaving the answers as exact values, complete with the square-root symbol, rather than decimals.

Although exact values are more precise, using decimal estimates of radical values is more helpful in practical situations. If you're solving for the height of a building and get $\sqrt{183}$, you get a better understanding of the height by finding a decimal estimate. A scientific calculator tells you that $\sqrt{183}$ is approximately 13.52774926. . . . Different calculators may give you fewer or more decimal values than I show you here. Usually, just two or three decimal places will do. Rounding this to two places, you get 13.53. Rounding it to three places, you get 13.528. (For more on rounding, please refer to Chapter 1.)

Getting to the Center of It All

One way to describe the middle of a triangle is to identify the *centroid*. This middle-point is the center of gravity, where you could balance the triangle and spin it around. And the middle of a line segment is its *midpoint*. When you graph a circle, triangle, or line segment by using coordinate axes, then you can name these middle points with a pair of *x*- and *y*-coordinates. All you need to find these middles are the coordinates of some crucial other points on the respective figures.

Finding the midpoint of a line segment

To find the midpoint of a line segment, you just calculate the averages of the coordinates — easy as pie.

The midpoint, *M*, of a segment with endpoints (x_1, y_1) and (x_2, y_2) is

$$M = \left(\frac{x_1 + x_2}{2}, \frac{y_1 + y_2}{2} \right)$$

If you want to know the midpoint of the segment with endpoints (–4,–1) and (2,5), then plug the numbers into the midpoint formula, and you get a midpoint of (–1,2):

$$M = \left(\frac{-4 + 2}{2}, \frac{-1 + 5}{2} \right) = \left(\frac{-2}{2}, \frac{4}{2} \right) = (-1, 2)$$

See how this segment looks in graph form in Figure 2-5.

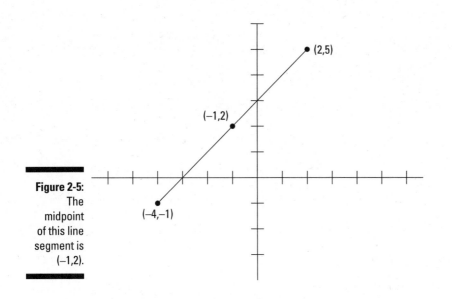

Locating the center of a circle

If the endpoints of one diameter of a circle are (x_1,y_1) and (x_2,y_2), then the center of the circle has the coordinates $\left(\dfrac{x_1+x_2}{2},\dfrac{y_1+y_2}{2}\right)$. You probably noticed that the center of a circle is the same as the diameter's midpoint. The center of the circle separates the diameter into two equal segments called *radii* (plural for radius).

Figure 2-6 shows a circle with a diameter whose endpoints are (7,4) and (–1,–2). The center of the circle is at (3,1). I got the coordinates for the center by using the formula for the midpoint of a segment (see the preceding section):

$$M = \left(\frac{7+(-1)}{2},\frac{4+(-2)}{2}\right) = \left(\frac{6}{2},\frac{2}{2}\right) = (3,1)$$

You find the length of the diameter by using the distance formula (see the section "Another slant: Diagonal distances," earlier in this chapter):

$$d = \sqrt{\left(7-(-1)\right)^2 + \left(4-(-2)\right)^2} = \sqrt{8^2 + 6^2} = \sqrt{64+36} = \sqrt{100} = 10$$

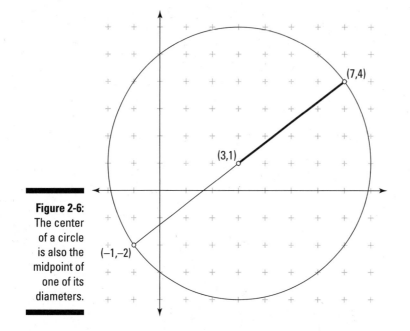

Figure 2-6:
The center
of a circle
is also the
midpoint of
one of its
diameters.

For the circle shown in Figure 2-6, the diameter is 10 units long.

Next, I show you how to find the length of one of the radii. Either will do — they're the same length. In this example, I figure the radius length from the center of the circle (3,1) to the endpoint of the diameter (7,4):

$$d = \sqrt{(7-3)^2 + (4-1)^2} = \sqrt{4^2 + 3^2} = \sqrt{16+9} = \sqrt{25} = 5$$

The radius is 5 units long. But, of course, you expected this answer, because by definition, the radius is half the length of the diameter.

Partitioning line segments further

If you can find the midpoint of a segment, you can divide it into two equal parts. Finding the middle of each of the two equal parts allows you to find the points needed to divide the entire segment into four equal parts. Finding the middle of each of these segments gives you eight equal parts, and so on.

For example, to divide the segment with endpoints (–15,10) and (9,2) into eight equal parts, find the various midpoints like so:

✔ The midpoint of the main segment from (–15,10) to (9,2) is (–3,6).

✔ The midpoint of half of the main segment, from (–15,10) to (–3,6), is (–9,8), and the midpoint of the other half of the main segment, from (–3,6) to (9,2), is (3,4).

✔ The midpoints of the four segments determined above are (–12,9), (–6,7), (0,5), and (6,3).

Figure 2-7 shows the coordinates of the points that divide this line segment into eight equal parts.

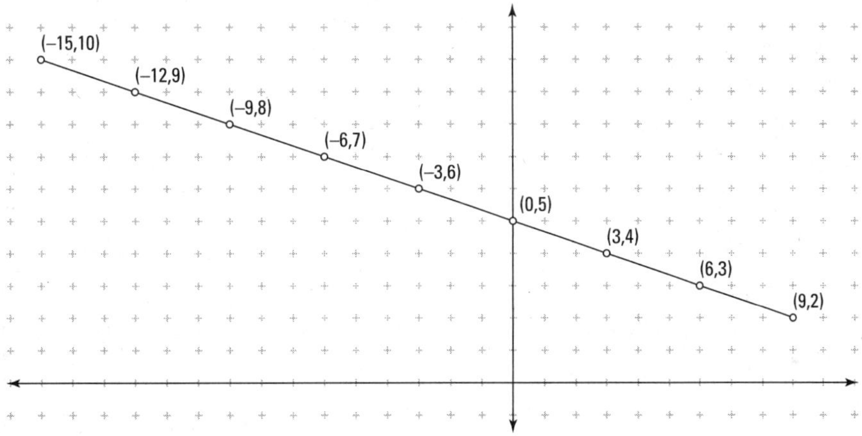

Figure 2-7:
A line
segment
divided into
eight equal
parts using
the midpoint
method.

Using the midpoint method is fine, as long as you just want to divide a segment into an even number of equal segments. But your job isn't always so easy. For instance, you may need to divide a segment into three equal parts, five equal parts, or some other odd number of equal parts.

To find a point that isn't equidistant from the endpoints of a segment, just use this formula:

$$(x,y) = \left(x_1 + k(x_2 - x_1), y_1 + k(y_2 - y_1)\right)$$

In this formula, (x_1, y_1) is the endpoint where you're starting, (x_2, y_2) is the other endpoint, and k is the fractional part of the segment you want.

So, to find the coordinates that divide the segment with endpoints (–4,1) and (8,7) into three equal parts, first find the point that's one-third of the distance from (–4,1) to the other endpoint, and then find the point that's two-thirds of the distance from (–4,1) to the other endpoint. The following steps show you how.

To find the point that's one-third of the distance from (–4,1) to the other endpoint, (8,7):

1. **Replace x_1 with –4, x_2 with 8, y_1 with 1, y_2 with 7, and k with $\frac{1}{3}$.**

   $$(x,y) = \left(-4 + \frac{1}{3}(8 - (-4)), 1 + \frac{1}{3}(7 - 1)\right)$$

2. **Subtract the values in the inner parentheses.**

   $$= \left(-4 + \frac{1}{3}(12), 1 + \frac{1}{3}(6)\right)$$

3. **Do the multiplication and then add the results to get the coordinates.**

   $$= (-4 + 4, 1 + 2) = (0,3)$$

To find the point that's two-thirds of the distance from (–4,1) to the other endpoint, (8,7):

1. **Replace x_1 with –4, x_2 with 8, y_1 with 1, y_2 with 7, and k with $\frac{2}{3}$.**

   $$(x,y) = \left(-4 + \frac{2}{3}(8 - (-4)), 1 + \frac{2}{3}(7 - 1)\right)$$

2. **Subtract the values in the inner parentheses.**

   $$= \left(-4 + \frac{2}{3}(12), 1 + \frac{2}{3}(6)\right)$$

3. **Do the multiplication and then add the results to get the coordinates.**

   $$= (-4 + 8, 1 + 4) = (4,5)$$

Figure 2-8 shows the graph of this line segment and the points that divide it into three equal parts.

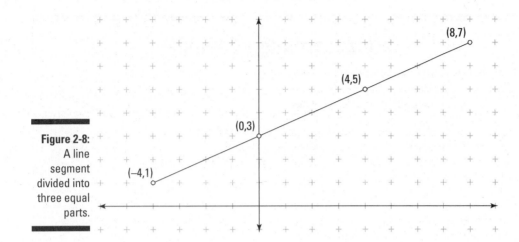

Figure 2-8:
A line segment divided into three equal parts.

Pinpointing the center of a triangle

If you draw lines from each corner (or *vertex*) of a triangle to the midpoint of the opposite sides, then those three lines meet at a center, or *centroid,* of the triangle. The centroid is the triangle's center of gravity, where the triangle balances evenly. The coordinates of the centroid are also two-thirds of the way from each vertex along that segment. Figure 2-9 shows how the three lines drawn in the triangle all meet at the center.

Figure 2-9:
The lines that intersect at the centroid of a triangle.

To find the centroid of a triangle, use the formula from the preceding section that locates a point two-thirds of the distance from the vertex to the midpoint of the opposite side.

Circumscribing a triangle

Every triangle can be *circumscribed* by a circle, meaning that one circle — and only one — goes through all three *vertices* (corners) of any triangle. In laymen's terms, any triangle can fit into some circle with all its corners touching the circle. To circumscribe a triangle, all you need to do is find the circumcenter of the circle (at the intersection of the perpendicular bisectors of the triangle's sides). You can then find the radius of the circle, because the distance from the center of the circle to one of the triangle's vertices is the radius. This exercise is a nice one to try your hand at with a compass and straightedge or with some geometry software.

For example, to find the centroid of a triangle with vertices at (0,0), (12,0) and (3,9), first find the midpoint of one of the sides. The most convenient side is the bottom, because it lies along the *x*-axis. The coordinates of that midpoint are (6,0). Then find the point that sits two-thirds of the way from the opposite vertex, (3,9):

1. **Replace x_1, x_2, y_1, and y_2 with their respective values. Replace k with $\frac{2}{3}$**

$$(x,y) = \left(3 + \tfrac{2}{3}\big(6 - (3)\big), 9 + \tfrac{2}{3}(0 - 9)\right)$$

2. **Simplify the computation to get the point.**

$$= \left(3 + \tfrac{2}{3}(3), 9 + \tfrac{2}{3}(-9)\right) = (3 + 2, 9 - 6) = (5,3)$$

In this example, the centroid is the point (5,3), as shown in Figure 2-10.

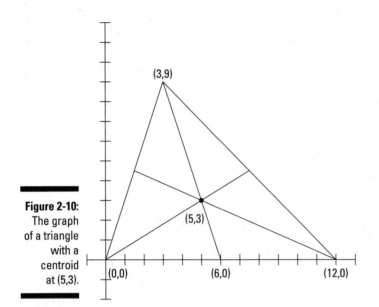

Figure 2-10: The graph of a triangle with a centroid at (5,3).

(3,9)

(5,3)

(0,0) (6,0) (12,0)

Racing Down the Slope

In mathematics, a *slope* is a particular number or value that tells you something about the nature of a line or line segment. Just by looking at the number corresponding to the slope of a line, you can tell if the line rises or falls as you read from left to right. You can also tell if the slope of the line is steep or rather flat (like the slopes in Colorado versus those in Illinois).

Slaloming slope formula

One way to find the slope of a line or segment is to choose any two points, (x_1,y_1) and (x_2,y_2), on the figure and use the formula that gives you the slope, represented by the letter m:

$$m = \frac{y_2 - y_1}{x_2 - x_1}$$

For example, the slope of the line through the points (–2,2) and (1,8) is

$$m = \frac{2-8}{-2-1} = \frac{-6}{-3} = 2$$

This line moves upward from left to right, which is why the slope is a positive number. Any slope greater than 1 is also considered to be *steep*.

On the other hand, the slope of the line through the points (–5,2) and (5,–1) is

$$m = \frac{2-(-1)}{-5-5} = \frac{3}{-10} = -\frac{3}{10}$$

This segment moves downward from left to right, so the slope is negative. The unsigned value (absolute value) of the slope is a number between 0 and 1 — so it isn't considered steep.

Figure 2-11 shows both lines, one with a slope of 2 and the other with a slope of $-\frac{3}{10}$.

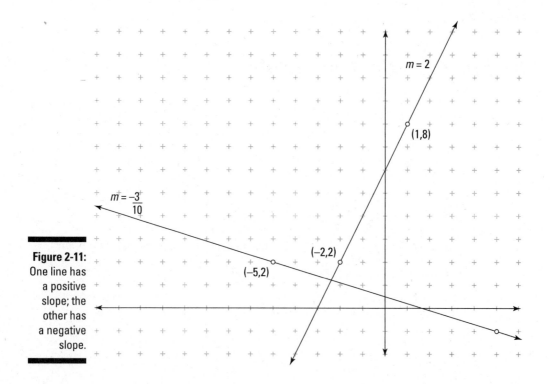

Figure 2-11:
One line has
a positive
slope; the
other has
a negative
slope.

Recognizing parallel and perpendicular lines

Two lines are *parallel* if they have the same slope. Two lines are *perpendicular* if their slopes are negative reciprocals of one another. Numbers that are *negative reciprocals* have a product –1.

Consider the following slopes of some lines or line segments:

$$m_1 = \frac{1}{2} \qquad\qquad m_2 = -2 \qquad\qquad m_3 = \frac{6}{3}$$

$$m_4 = \frac{5}{10} \qquad\qquad m_5 = -\frac{14}{7}$$

Here are the slopes of the lines that are parallel:

- ✔ $m_1 = \frac{1}{2}$ and $m_4 = \frac{5}{10}$ have the same slope.
- ✔ $m_2 = -2$ and $m_5 = -\frac{14}{7}$ also have the same slope.

Here are the slopes of the lines that are perpendicular:

- ✔ $m_1 = \frac{1}{2}$ and $m_2 = -2$ have slopes that are negative reciprocals.
- ✔ $m_4 = \frac{5}{10}$ and $m_5 = -\frac{14}{7}$ also have slopes whose product is –1.

As a matter of fact, because the lines with slopes of $\frac{1}{2}$ and $\frac{5}{10}$ are equal to one another, they're both perpendicular to the lines with slopes of –2 and $-\frac{14}{7}$, which are also equal in slope. It's one big, happy family.

Defining Circles with Numbers

The circle that you use the most in trigonometry has its center at the origin and has a radius of 1 unit (called the *unit circle*). The radius of 1 in a circle makes computations so much easier when that 1 ends up in the denominator of a fraction. Fractions and circles sort of intermingle in trigonometry — in good ways, of course. But you also have many other useful circles to consider. The other circles will have different radii and different centers, but each has its place when needed. When possible, though, the *unit circle* is the circle of choice.

Centering circles at the origin

The two characteristics that define a circle are its center and its radius. The center tells where on a graph the circle is located; the radius tells how big the circle is. The location is in terms of coordinates in the coordinate plane, and those numbers end up in the equation of the circle. The x and y variables represent the coordinates of all the points that lie on the actual circle. The standard form for the equation of a circle at the origin is $x^2 + y^2 = r^2$, where r represents the radius of the circle. So the equations for circles with radii of 2, 3, 4, and 5 are $x^2 + y^2 = 4$, $x^2 + y^2 = 9$, $x^2 + y^2 = 16$, and $x^2 + y^2 = 25$, respectively.

Likewise, a circle with a radius of 1 unit has the equation $x^2 + y^2 = 1$. This *unit circle* is used extensively in mathematics: The radius of 1 lends itself to the formula for changing from degrees to radians, is nice and neat when finding arc length, and makes the unit circle the easiest to use when proving properties or theorems in math.

Wandering centers

Circles don't have to have their centers at the origin. The standard form for a circle with a radius of r and its center at (h,k) is $(x - h)^2 + (y - k)^2 = r^2$, where x and y represent the coordinates of all the points on that circle. So, the equation for a circle with its center at $(3,-2)$ and with a radius of 9 is $(x - 3)^2 + (y + 2)^2 = 81$.

Notice that if you let the center of a circle be $(0,0)$ in this formula, you get $(x - 0)^2 + (y - 0)^2 = r^2$ or $x^2 + y^2 = 1$, which goes back to a circle with its center at the origin. The form works for all circles!

Chapter 3

Functioning Well

In This Chapter

▶ Understanding why functions are your friends

▶ Applying the inverse to a function

▶ Moving a function around on a graph

*Y*ou can't get very far in any mathematical discussion without encountering rules, patterns, operations, or relationships among the concepts you're discussing. One common theme in math is the relationship between certain values (often called the *input* and the *output*), which are the values you start with and the values you end up with, respectively. *Functions* are very special types of relationships using input and output values, and they play a big part in trigonometry. So, what distinguishes a relation from a function, and why should you care? The distinction is important in all mathematics, not just in trigonometry.

Relations versus Functions

A *relation* in mathematics is a rule that creates a certain output for any given input. The *input* is the number you enter in place of a variable, and the *output* is the result(s) you get when you perform the operations for that relation. Each relation has a rule, or expression, that usually involves mathematical operations such as addition, subtraction, square roots, and so on. For instance, you could come across a relation where you input 25 in place of a variable and get two output values, such as 24 and 27. The rule for that relation could be that you input a number and get the two numbers closest to it that are multiples of 3. This relation has more than one output value, which isn't necessarily a good thing; in math, more isn't always better.

A *function,* however, is a special kind of relation. Read on to find out more.

Function junction, what's your function?

A function in mathematics is a rule performing operations and processes on input values that results in a single, unique output value — only one output for every input. For example, take the function where the input is the number 25 and the output is 5. Now you have several ways of getting the result of 5 after inputting 25 in place of a variable: You can take the square root of the number or subtract 20 from the number, for starters. But the main emphasis here is that the function has just *one* answer or output value.

Consider a function that uses a *radical* (a root). Input 25, and here's what that function looks like: $\sqrt{25}$. The output is the single value 5. Another function is one that squares the input, multiplies that result by 2, and then subtracts 3; this can be written $2x^2 - 3$. If you input an 8, then you get $2(8)^2 - 3 = 2(64) - 3 = 128 - 3 = 125$. The preceding examples of functions use the basic algebraic operations. But I'm here to tell you that a whole class of functions called *trigonometric functions* is out there, too. That's why you're reading this book! One of the trig functions is called *sine*, abbreviated *sin*. If you compute the sine of 30 degrees, you get $\sin 30 = \frac{1}{2}$. Because sine is a function, $\frac{1}{2}$ is the only output value. It may seem trivial or unnecessary for me to keep harping on the *only one output* business right now, but having only one output value for each input in trigonometry is very important — otherwise, you'd have chaos!

Using function notation

Defining a function or explaining how it works can involve a lot of words and can get rather lengthy and awkward. Imagine having to write, "Square the input, multiply that result by 2, and then subtract 3." Mathematicians are an efficient lot, and they prefer a more precise, quicker way of writing their instructions. Function notation is just that.

First, functions are generally named with letters — the most frequently used is the oh-so-obvious *f*. (I said that mathematicians are efficient, but they're not necessarily original or creative.) If I want the function *f* to be the rule for squaring a number, multiplying that result by 2, and then subtracting 3, I write the function as $f(x) = 2x^2 - 3$. You read the function like this: "*f* of *x* is equal to two times *x* squared minus 3." The *x* is a variable — in this case, the input variable. Whatever you put in the parentheses after the *f* replaces any *x* in the rule. In the first equation that follows, an 8 replaces the *x*. In the second equation, a –4 replaces the *x*. Each time, the function produces only one answer:

$$f(8) = 2(8)^2 - 3 = 128 - 3 = 125$$

$$f(-4) = 2(-4)^2 - 3 = 32 - 3 = 29$$

Don't feel bound to the f, though. You can use other letters to name the functions and the input variables. Sometimes you use letters that represent what's going on or what you're using the formula for, such as finding area, interest, or cost:

$$A(r) = \pi r^2$$

$$I(t) = 1,000\, e^{0.04t}$$

$$C(x) = -0.04x^2 + 8x + 100$$

And, of course, you have the trig functions. Some trig functions involving sine, cosine, and secant are

$$p(x) = \sin x + \cos x$$

$$c(\theta) = \frac{1}{\sec \theta}$$

Determining domain and range

A function consists of a rule that you apply to the input values. The result is a single output value. You can usually use a huge number of input values, and they're all part of the *domain* of the function. The output values make up the *range* of the function.

Are you master of your domain?

The domain of a function consists of all the values that you can use as input into the function rule. The domain is another of that function's characteristics, because different functions have different numbers that you can input and have the outputs make any sense.

For example, $f(x) = \sqrt{x}$ is a function whose domain can't contain any negative numbers, because the square root of a negative number isn't a real number.

The function $g(x) = \frac{4}{x+3}$ has a domain that can't include the number -3. Any other real number is okay, but not -3, because putting a -3 in for x makes the denominator equal to 0, and you can't divide by 0. (A fraction with a 0 in the denominator represents a number that doesn't exist.) With trig functions, the domain (input value) is angle measures — either in degrees or radians. Some of the trig functions have restrictions on their domains, too. For example, the tangent function has a domain that can't include 90 degrees or 270 degrees, among the many other restricted values. (I discuss these domains in detail in Chapter 7.)

Home, home on the range

The range of a function consists of all its output values — the numbers you get when you input numbers from the domain into the function and perform the function operations on them. Sometimes, a range can be all possible real numbers — it has no limit. That situation happens in a function such as $h(x) = 3x + 2$. In this equation, both the domain and the range are unlimited. You can put in any real number, and you can get an output of any real number that you can possibly think of. Ranges can end up being restricted, though. For example, the function $k(x) = x^2 + 6$ will always have results that are either the number 6 or some positive number greater than 6. You can never get a negative number or a number less than 6 as an output. The ranges of some trig functions are restricted, too. For example, the output of the sine function never exceeds 1 or goes lower than –1. (I cover this subject in detail in Chapter 7.)

In-Verse Functions: Rhyme or Reason?

Functions are special types of relationships between mathematical values, because they yield only one unique output value for every input value. (For a more-detailed definition of a function, see the "Function junction, what's your function?" section, earlier in this chapter.) Sometimes, you have to work backward with functions, because you know the output value and you want to figure out which input value gave you that output. That's where *inverse functions* come in.

Which functions have inverses?

The best way to describe an inverse function is to give an example. I show you two functions: One formula or function tells you what the temperature outside is in degrees Celsius when you input a temperature in degrees Fahrenheit, and the other gives you the degrees Fahrenheit when you input Celsius (very handy when traveling abroad).

The first function is $C(f) = \frac{5}{9}(f - 32)$, and the second is $F(c) = \frac{9}{5}c + 32$, where C is the answer in Celsius degrees when you input f as the temperature in Fahrenheit degrees, and then F is the answer in Fahrenheit when you input c as the temperature in Celsius.

If you input 77 degrees Fahrenheit into the function C, you get $C(77) = \frac{5}{9}(77 - 32) = \frac{5}{9}(45) = 25$, or 25 degrees Celsius.

Now, how about going the other way? What if you want to know what temperature in Celsius will *give you* 77 degrees Fahrenheit? You use the inverse of the function (which I show you how to find in the next section). Keep in mind that only one temperature gives you that answer of 77 degrees. (It wouldn't make sense if both 25 degrees and 45 degrees Celsius, for example, gave the same answer of 77 degrees Fahrenheit.)

Using the other function and inputting 25 degrees Celsius,
$$F(25) = \frac{9}{5}(25) + 32 = 45 + 32 = 77.$$

There is a big distinction between functions that have inverses and functions that don't: A function can have an inverse function only if the function is *one-to-one*. In other words, the function has to be designed in such a way that every input has exactly one output *and* every output comes from only one input — the output doesn't occur with more than one input value.

An example of another function that has an inverse function is $f(x) = 4x + 5$. Its inverse is $f^{-1}(x) = \frac{x-5}{4}$.

Notice that the function notation for the name of the inverse function is the same letter but with a −1 exponent. This exponent doesn't mean that you want a reciprocal; instead, the −1 exponent in a function name is special math notation meaning an inverse function. You see this notation a lot in trigonometry. The inverse functions all have a name using the −1 exponent after the corresponding function name. They can have an alternative name, too. For more on this naming mumbo-jumbo, go to Chapter 15.

Check out how this inverse function works by using the last function I showed you. If you input a 6 into the function, f, you get $f(6) = 4(6) + 5 = 24 + 5 = 29$. Take that output, 29, and put it into the inverse function to see where that particular output came from: $f^{-1}(29) = \frac{29-5}{4} = \frac{24}{4} = 6$. This function and its inverse are one-to-one. No other input into $f(x)$ will give you 29, and no other input into $f^{-1}(x)$ will give you 6.

Not all functions have inverses, though. An example of a function that is *not* one-to-one is $g(x) = x^2 - 4$. If you input 7, you get $g(7) = (7)^2 - 4 = 49 - 4 = 45$. But you also get 45 if you input −7: $g(-7) = (-7)^2 - 4 = 49 - 4 = 45$. From the output value, you can't possibly tell which of the two numbers was the input value, the 7 or −7. This function isn't one-to-one, so it doesn't have an inverse.

Sometimes, you can spot a function that has an inverse, and sometimes that quality isn't so apparent. Here are some fairly obvious clues that you can pick up on just by looking at the rule for the function. A function does *not* have an inverse if

    ✔ The function rule has an even exponent (not including 0).

    ✔ The function rule has an absolute value symbol.

    ✔ The graph of the function is a horizontal line.

    ✔ You draw a horizontal line through the graph of the function and that line intersects the graph more than once.

The trig functions all have inverses, but only under special conditions — you have to restrict the domain values. (I discuss what it means to restrict the domain values in Chapter 15.)

Finding an inverse function

Not all functions have inverses, and not all inverses are easy to determine. Here's a nice method for finding inverses of basic algebraic functions.

Using algebra

The most efficient method for finding an inverse function for a given one-to-one function involves the following steps:

    1. **Replace the function notation name with *y*.**

    2. **Reverse all the *x*'s and *y*'s (let every *x* be *y* and every *y* be *x*).**

    3. **Solve the equation for *y*.**

    4. **Replace *y* with the function notation for an inverse function.**

For example, to find the inverse function for $f(x) = \sqrt[3]{x-2} + 8$:

    1. **Replace the function notation with *y*.**

$$y = \sqrt[3]{x-2} + 8$$

    2. **Reverse the *x*'s and *y*'s.**

$$x = \sqrt[3]{y-2} + 8$$

    3. **Solve for *y*.**

$$x - 8 = \sqrt[3]{y-2}$$
$$(x-8)^3 = y - 2$$
$$(x-8)^3 + 2 = y$$

4. **Replace _y_ with the inverse function notation.**

$$f^{-1}(x) = (x-8)^3 + 2$$

Look at how these two functions work. Input 3 into the original function and then get the number 3 back again by putting the output, 9, into the inverse function.

1. **Replace the _x_'s with 3 in the function.**

$$f(3) = \sqrt[3]{3-2} + 8 = \sqrt[3]{1} + 8 = 9$$

2. **Replace the _x_'s with 9 in the inverse function.**

$$f^{-1}(9) = (9-8)^3 + 2 = 1^3 + 2 = 3$$

Using new definitions of functions for inverses

Sometimes you just don't have a nice or convenient algebraic process that will give you an inverse function. Many functions need a special, new rule for their inverse. Here are some examples of these functions:

Function	Inverse
$f(x) = e^x$	$f^{-1}(x) = \ln x$
$g(x) = \log_a x$	$g^{-1}(x) = a^x$
$h(x) = \sin x$	$h^{-1}(x) = \arcsin x$ or $\sin^{-1}x$
$k(x) = \tan x$	$k^{-1}(x) = \arctan x$ or $\tan^{-1}x$

If you have a scientific or graphing calculator, you can try out some of these functions and their inverses. Use the function $f(x) = e^x$ and its inverse, $f^{-1}(x) = \ln x$, for the following demonstration:

1. **In the calculator, use the _e^x_ button (often a second function of the calculators) to enter _e³_.**

   The input value here is 3. The answer, or output, comes out to be about 20.08553692. This value isn't exact, but it's good for eight decimal places.

2. **Now take that answer and use the _ln_ button to find _ln_ 20.08553692.**

   Input 20.08553692 into the _ln_ function. The answer, or output, that you get this time is 3.

Transforming Functions

Functions and all their properties, characteristics, and peculiarities are of interest to mathematicians and others who use them as models for practical applications. Using functions to find values or answers to practical problems is helpful only if tweaking or slightly changing the functions is reasonably simple. Predictable and controlled changes of functions meet this requirement of ease and simplicity. Chapter 22 deals with how the transformations affect trig functions. This section gives you a more-general explanation of how to tweak your functions.

Translating a function

A *translation* is a slide, which means that the function has the same shape graphically, but the graph of the function slides up or down or slides left or right to a different position on the coordinate plane.

Sliding up or down

Figure 3-1 shows the parabola $y = x^2$ with a translation 5 units up and a translation 7 units down. A *parabola* is the graph of a second-degree polynomial, which means that the polynomial has a power of 2 for one exponent. The graph makes a nice, U-shaped curve.

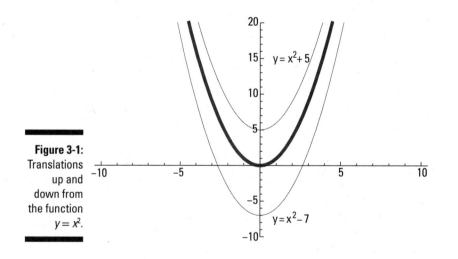

Figure 3-1: Translations up and down from the function $y = x^2$.

Think about a function that you use to determine how much money a person earns for working a certain number of hours. The amount can slide up or down if you add a bonus or subtract a penalty from the amount. Here's what the situation may look like in function notation:

> ✔ **Translating up _C_ units:** $f(x) + C$
>
> ✔ **Translating down _C_ units:** $f(x) - C$

A person who makes $8 an hour but gets a $50 bonus has a pay function for _h_ hours that looks like $P(h) = 8h + 50$. If that same person were penalized $6 for being late, the pay function would look like $P(h) = 8h - 6$.

Sliding left or right

Figure 3-2 shows the parabola $y = x^2$ with a translation 5 units right and a translation 7 units left.

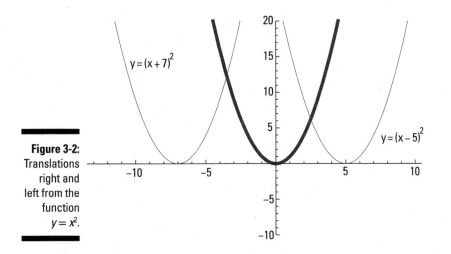

Figure 3-2: Translations right and left from the function $y = x^2$.

If you use a function to determine how much commission a person earns for selling a certain number of computers, the commission can be affected when you add or subtract the number of units the person needs to sell. Here's what the situation looks like in function notation:

> ✔ **Translating left _C_ units:** $f(x + C)$
>
> ✔ **Translating right _C_ units:** $f(x - C)$

A person who makes $50 commission for every computer sold but gets upfront credit for two computers as an incentive has a commission function for _x_ computers that looks like $P(x) = 50(x + 2)$. On the other hand, a person who has the same commission schedule but had two computers returned and starts with a deficit has a commission function that looks like $P(x) = 50(x - 2)$.

Reflecting like a mirror

Two types of transformations act like reflections or flips. One transformation changes all positive outputs to negative and all negative outputs to positive. The other reverses the inputs — positive to negative and negative to positive.

- ✔ **Reflecting up and down (outputs changed):** $-f(x)$
- ✔ **Reflecting left and right (inputs changed):** $f(-x)$

Figure 3-3 shows reflections of the function $y = \sqrt{x}$. Reflecting downward puts all the points below the *x*-axis. Reflecting left makes all the input values move to the left of the *y*-axis. Even though it appears that the negatives shouldn't go under the radical, in fact, the negative in front of the *x* means that you take the opposite of all the negative *x*'s — which makes them positive.

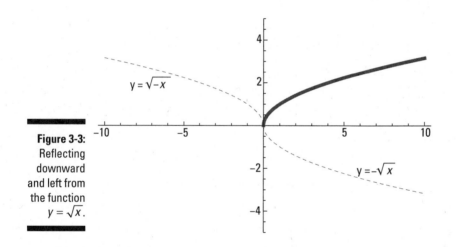

Figure 3-3: Reflecting downward and left from the function $y = \sqrt{x}$.

A cash register can change inputs to the opposite (negative) numbers by taking coupon values that the cashier enters or scans in and changing them to negative values before doing the final computations. The graph of this process acts as a reflection downward from positive to negative.

Left and right reflections are a bit harder to describe in terms of a practical application. Try this one on for size: If a function tells you how many items a machine can produce in a certain number of hours, then inputting negative numbers helps you determine how far you have to back up — how many hours before a certain date and time — to produce that number of items by that date and time.

Chapter 4

Getting Your Degree

In This Chapter

▶ Measuring angles in degrees

▶ Putting angles in standard position

▶ Finding many measures for the same angle

*T*he main idea that distinguishes trigonometry from other mathematical topics is its attention to and dependence on angle measures. The trig functions (sine, cosine, tangent, cotangent, secant, and cosecant) are ratios based on the measures of an angle. What good are degrees (no, not the kind that tell you how hot or cold it is) in the real world? Navigators, carpenters, and astronomers can't do without them. How do you measure the degrees? You have many ways, dear reader, and I show you all you need to know in this chapter.

Angles, Angles Everywhere: Measuring in Degrees

What's a degree? When you graduate from college, you get your degree. The temperature outside went up a degree. When questioned, you get the third degree. All these scenarios use the word *degree,* but in trigonometry, a degree is a tiny slice of a circle. Imagine a pizza cut into 360 equal pieces (what a mess). Each little slice represents one degree. Look at Figure 4-1 to see what a degree looks like.

Slicing a coordinate plane

The first quadrant is the upper right-hand corner of the coordinate plane. (See Chapter 2 for the lowdown on quadrants and coordinate planes.) That first quadrant is $\frac{1}{4}$ of the entire plane. So, if a full circle with its center at the origin has a total of 360 degrees, then $\frac{1}{4}$ of it has 90 degrees, which is the measure of the angle that the first quadrant forms. Actually, each quadrant measures exactly 90 degrees. You can divide each of these 90-degree measures evenly by many numbers, and you use those equal divisions frequently in trig, because they're nice, neat divisions. The most frequently used angle measures include $\frac{90}{2} = 45°$, $\frac{90}{3} = 30°$, and $\frac{90}{6} = 15°$. And then, twice the 30-degree angle is 60 degrees (another common angle in trig).

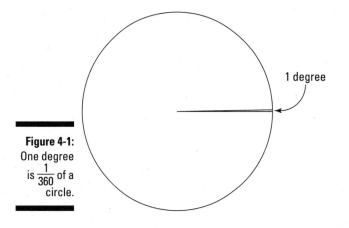

Figure 4-1:
One degree
is $\frac{1}{360}$ of a
circle.

This elite group of angle measures is 0, 15, 30, 45, 60, and 90 degrees. These angles and their multiples occupy much of the discussion in trigonometry because of their convenience in computations. Figure 4-2 shows sketches of some of the angles.

Figure 4-2:
Some of
the most
commonly
used angles:
90, 60, 45,
30, and 15
degrees.

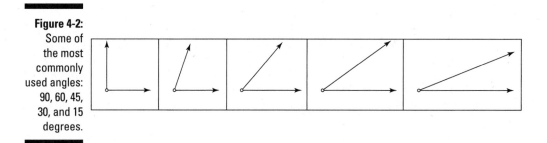

Looking elsewhere for degree measures

Your first introduction to the idea of measuring angles in degrees probably didn't come from a course in geometry or trigonometry. Most of us are exposed to this idea through television or movies. A popular situation in such shows involves a plane flying through a storm or at night or with no one but a stewardess at the controls. A radio transmission from the control tower comes crackling through all the static, with an announcer saying, "Turn to a heading of 40 degrees." And because the pilot or stewardess remembers her trigonometry, she saves the day. Hurray for degrees!

Another type of situation that you find on television is on *This Old House*, where the stars, in all their woodworking grandeur, are able to cut boards at exact 50-degree angles so they fit perfectly in a carefully crafted wooden truss.

Navigating with degrees

In navigation and surveying, the *bearing* or *heading* is the direction that a plane, boat, or line takes. In math-speak, this bearing is the angle measured in degrees that a *ray* (a line with one endpoint that extends infinitely in the other direction) makes with a second ray that points north. The angle is measured in a clockwise direction. (Note, however, that in the *standard position* in geometry and trigonometry, you measure angles in a *counterclockwise* direction.) Figure 4-3 shows some bearings used in navigation. Notice that the direction of the arrow is always clockwise. Even though the angles in bearings are measured differently from those in trigonometry, the angle measures are still the same size — just rotated a bit. An angle of 120 degrees is still bigger than a right angle. When you're familiar with the angle sizes, translating into this bearing business is easy.

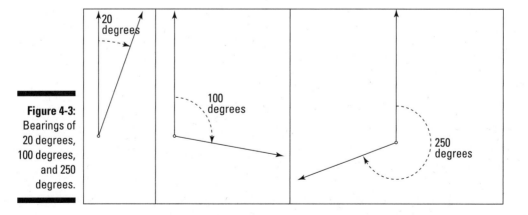

Figure 4-3:
Bearings of
20 degrees,
100 degrees,
and 250
degrees.

Columbus the wizard

It's a given that trigonometry played a big part in navigation and allowed Christopher Columbus to find the New World. But trigonometry also helped him in another way. On his voyages, Columbus carried a copy of an almanac created by a mathematician/astronomer by the name of Johannes Müller. In the almanac were tables giving the relative positions of the sun and moon and which determined when and where eclipses would occur. Columbus read that a total eclipse of the moon would occur on February 29, 1504. He took advantage of this information and used it to frighten the natives in the New World into supplying provisions for his ships.

Now, take a look at Figure 4-4 to see the path of a helicopter pilot who flew for 10.5 minutes at a bearing of 36 degrees (which is northeast), then for 13.6 minutes at a bearing of 144 degrees (which is southeast), and then got back to where she started by flying for 14.4 minutes at a bearing of 280 degrees (which is west-northwest).

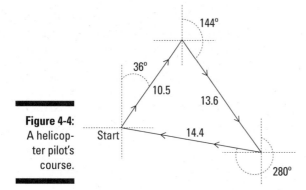

Figure 4-4:
A helicopter pilot's course.

Understanding Norm's workshop

If you aren't a follower of public television and *The New Yankee Workshop* starring Norm Abram, let me fill you in: Norm is a New Englander who does woodworking projects with very expensive tools and invites his audience to do the same (with not-so-expensive tools). He sets his table saw so it can cut a board straight across at a 90-degree angle, or he changes it to cut at any other angle. If Norm wants two perpendicular pieces of wood to meet and form a right angle, he sets his saw at 45 degrees. He can also cut one piece at 30 degrees and the other at 60 degrees; or how about 20 and 70? Figure 4-5 shows how the two pieces of wood fit together.

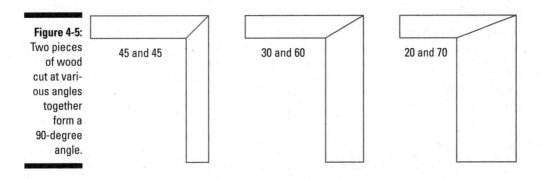

Figure 4-5:
Two pieces of wood cut at various angles together form a 90-degree angle.

45 and 45   30 and 60   20 and 70

If Norm wants to create an octagonal (eight-sided) table from a single piece of wood that he cuts into eight pieces, then what angles should he cut? More on that in a minute. In Figure 4-6, you see an octagonal table constructed of eight equal triangles.

Figure 4-6:
An octagonal table and one of the pieces that comprises it.

To make his octagonal table, Norm needs eight *isosceles triangles* (where the two long sides of each triangle are the same length). What are the measures of the angles he has to cut? All the way around a circle (and around the middle of the table) is 360 degrees, so each triangle has a *top angle* (the angle at the center of the table) that measures $\frac{360}{8}$, which is 45 degrees. The two *base angles* (those at the outer edge of the table) are equal in measure. The sum of the measures of the angles of a triangle is 180 degrees, so after subtracting the top angle's 45 degrees, you get 135 degrees for the other two angles together. Dividing the 135 by 2, you find that the base angles are each $67\frac{1}{2}$ degrees. Norm can cut all eight triangles from a single piece of wood, because two base angles plus a top angle form a straight line. He'll just put the triangles together differently after cutting them out. And as you can see from Figure 4-7, he doesn't have much waste.

Figure 4-7:
Cutting eight
identical
triangles out
of a board.

Graphing Angles in Standard Position

Navigators, surveyors, and carpenters all use the same angle measures, but the angles start out in different positions or places. In trigonometry and most other mathematical disciplines, you draw angles in a standard, universal position, so that mathematicians around the world are drawing and talking about the same thing.

Positioning initial and terminal sides

An angle in *standard position* has its vertex at the origin of the coordinate plane, as shown in Figure 4-8. Its *initial* ray (starting side) lies along the positive *x*-axis. Its *terminal ray* (ending side) moves counterclockwise from the initial side.

If the terminal ray moves clockwise instead of counterclockwise, then the measure is a negative value. You often name angles in standard position with a Greek letter.

The lengths of the rays that create the angle have nothing to do with the angle size. You can extend rays as long as you need them to be, and the angle measure won't change. Only the position of the terminal ray determines the angle.

Measuring by quadrants

Angles in the standard position are used in calculus, geometry, trigonometry, and other math subjects as a basis for discussion. Being able to recognize a particular angle by the quadrant its terminal side lies in and, conversely, to know which angles have their terminal sides in a particular quadrant is helpful when working in these areas.

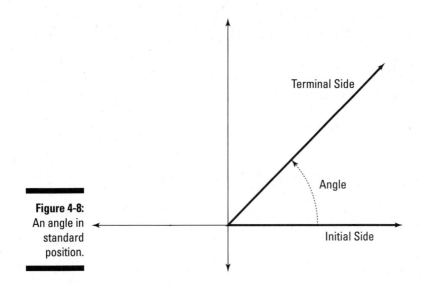

Figure 4-8:
An angle in
standard
position.

Check out Figure 4-9. Angles in standard position that measure between 0 and 90 degrees have their terminal sides in Quadrant I. The angles measuring between 90 and 180 degrees have their terminal sides in Quadrant II. Angles measuring between 180 and 270 have their terminal sides in Quadrant III, and those measuring between 270 and 360 have their terminal sides in Quadrant IV. Angles measuring exactly 90, 180, 270, and 360 degrees do not have a terminal side that lies in a quadrant, and they're referred to as *quadrant angles*.

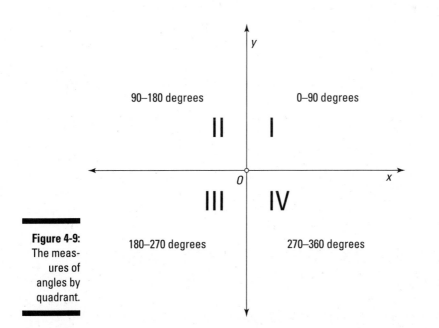

Figure 4-9:
The meas-
ures of
angles by
quadrant.

What's Your Angle? Labeling in Various Ways

The terminal side of an angle determines its angle measure. But more than one angle has the same terminal side — in fact, an infinite number of angles share a particular terminal side.

Using negative angle measures

If you want your angle measurement to be positive, you measure the angle in standard position in a *counterclockwise* direction. However, angles can have negative values, too, as you see in Figure 4-10. You get a negative value when you measure an angle in a *clockwise* direction. Therefore, an angle of 300 degrees has the same terminal side as an angle measuring –60 degrees. If they have the same terminal side, then why don't they have the same name/size? And which name is better? Sometimes you may want to keep the numerical part of the measure smaller. For example, picturing an angle of –30 degrees is easier than picturing one of 330 degrees. Also, pilots don't always have the choice as to which direction they can turn in, but going 10 degrees in the negative direction makes more sense than going 350 degrees — all the way around, practically — in the positive direction. One common practice is to name all angles with a number that has an absolute value less than 180 degrees. So –60 degrees is often preferable to 300 degrees.

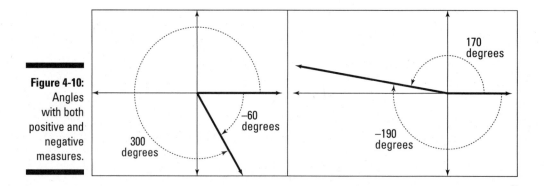

Figure 4-10: Angles with both positive and negative measures.

Comingling with coterminal angles

Two angles are *coterminal* if they have the same terminal side. You have an infinite number of ways to give an angle measure for a particular terminal ray. Sometimes, using a negative angle rather than a positive angle is more convenient, or the answer to an application may involve more than one revolution (spinning around and around). Angles can have terminal sides that involve one or more full revolutions around the origin or terminal sides that go clockwise instead of counterclockwise — or both of these situations can happen.

More than one revolution

An angle measuring 70 degrees is coterminal with an angle measuring 430 degrees (see Figure 4-11). The angle measuring 430 degrees is actually 360 + 70 (one full revolution plus the original 70). These two angles are also coterminal with an angle of 790 degrees (360 + 360 + 70 = 790). This pattern could go on and on, with the addition of another 360 degrees each time.

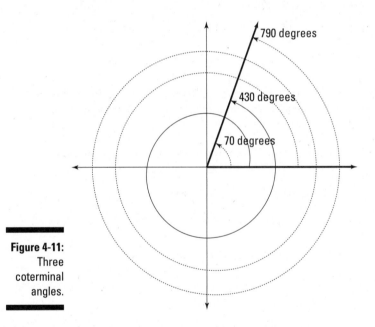

Figure 4-11:
Three
coterminal
angles.

Negative coterminal angles

An angle of 70 degrees is coterminal with an angle of –290 degrees. Two rotations in the negative (clockwise) direction give you an angle of –650 degrees (–290 – 360 = –650).

Renaming angles: So many aliases

Any angle can have many, many descriptions in terms of angle measures, because an angle is equivalent to its coterminal angles. The most frequently used positive angle measures are those that measure between 0 and 360 degrees.

Rules for coterminal angles involve adding or subtracting *rotations* (or multiples of 360 degrees). The first equation that follows shows what happens when you add a full rotation over and over. The second shows what happens when you subtract a full rotation many times. The results are all coterminal angles.

$$\theta \to \theta + 360° \to \theta + 720° \to \theta + 1,080° \to ... \to \theta + 360 \cdot k°$$

$$\theta \to \theta - 360° \to \theta - 720° \to \theta - 1,080° \to ... \to \theta - 360 \cdot k°$$

So an angle measuring 100 degrees is coterminal with the following:

Adding: $100° \to 100° + 360° \to 100° + 720° \to 100° + 1,080° \to ... \to 100° + 360 \cdot k°$

$100° \to 460° \to 820° \to 1,180°$

Subtracting: $100° \to 100° - 360° \to 100° - 720° \to 100° - 1,080° \to ... \to 100° - 360 \cdot k°$

$100° \to -260° \to -620° \to -980°$

Here's an example: Suppose you want to give new measures for angles of 800 degrees and –1,040 degrees by finding an equivalent angle measure between 0 and 360 degrees.

1. **Subtract 360 degrees from 800 until the result is less than 360.**

   $800° \to 800° - 360° = 440°$

   $440° \to 440° - 360° = 80°$

   An angle measuring 800 degrees is coterminal with an angle of 80 degrees.

2. **Add 360 degrees to –1,040 until the result is positive.**

   $-1,040° \to -1,040° + 360° = -680°$

   $-680° \to -680° + 360° = -320°$

   $-320° \to -320° + 360° = 40°$

   An angle measuring –1,040 degrees is coterminal with an angle of 40 degrees.

Chapter 5

Dishing Out the Pi: Radians

A person's first introduction to angles is usually in terms of degrees. You probably have an idea of what a 30-degree angle looks like. (If not, review Chapter 4.) And even most middle-school students know that a triangle consists of 180 degrees. But most of the scientific community uses radians to measure angles and solve trig equations. Why change to radians? Why fix what ain't broke? Read on.

What's in a Radian?

A radian is much bigger than a degree. Early mathematicians decided on the size of a degree based on divisions of a full circle. A degree is the same as a slice of $\frac{1}{360}$ of a circle. No one knows for sure how the choice of 360 degrees in a circle came to be adopted. In any case, 360 is a wonderful number, because you can divide it evenly by so many other numbers: 2, 3, 4, 5, 6, 8, 9, 10, 12, 15, 18, 20, 24, 30, 36, 40, 45, 60, 72, 90, 120, 180, and 360. The early measures of time and distance relied on having convenient numbers to work with. A radian, on the other hand, isn't quite as nice. It isn't even a rational number. Radians probably were developed because mathematicians wanted to relate the angle measure more to the radius or size of the circle. A circle has 2π radians (a little more than six radians). A radian is *almost* $\frac{1}{6}$ of a circle — it's a little more than 57 degrees. Figure 5-1 compares a degree with a radian.

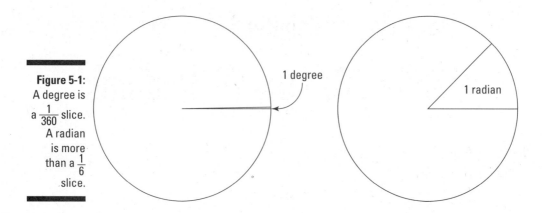

Figure 5-1:
A degree is a $\frac{1}{360}$ slice. A radian is more than a $\frac{1}{6}$ slice.

Relating to a circle

The big advantage of using radians is that they're the *natural* measure for dividing up circles. Imagine taking the radius of a circle and bending it into an arc that lies along the circle. Now draw radii from the center to both ends of that arc formed by the radius. The angle formed from the radii measures one radian. You would need a little more than six of those arcs to go all the way around the circle. This fact is true of all circles. The circumference of any circle is always a little more than three times the diameter of that circle — π times the diameter, to be exact. Another way of saying this is 2π times the radius. That number may seem nicer and more civilized than the big number 360, but the disadvantage is that π doesn't have an exact decimal value. Saying 2π radians (which is equal to 360 degrees) means that each circle has *about* 6.28 radians. Even though radians are the natural measure and always relate to the radius and diameter, the decimal values get a bit messy.

Each of these measures has its own place. Measuring angles in degrees is easier, but measuring angles in radians is preferable when doing computations. The radian is more exact because the radius, circumference, or area of the circle is involved. Even though π doesn't have an exact decimal value, when you use multiples of π in answers, they're exactly right. I show you an example of using π as part of an answer in "Making a Clone of Arc," later in this chapter.

Converting degrees and radians

Many math problems require changing from degrees to radian measures or vice versa. You often perform mathematical computations in radians, but then convert the final answers to degrees so the answers are easier to visualize and comprehend. You can use a nifty little proportion to change from degrees to radians or radians to degrees. In this proportion, the Greek letter theta, θ, represents the name of the angle. Putting the superscripts $\circ$ and R on θ makes the angle stand for the measure in degrees and radians, respectively.

$$\frac{\theta^\circ}{180} = \frac{\theta^R}{\pi}$$

This proportion reads: "The measure of angle θ in degrees divided by 180 is equal to the measure of angle θ in radians divided by π." (Remember that π is about 3.141592654.)

The computation required for changing degrees to radians and radians to degrees isn't difficult. The computation involves a few tricks, though, and the format is important. You don't usually write the radian measures with decimal values unless you've multiplied through by the decimal equivalent for π.

Changing degrees to radians

To change a measure in degrees to radians, start with the basic proportion for the equivalent angle measures: $\frac{\theta^\circ}{180} = \frac{\theta^R}{\pi}$.

For example, here's how you change a measure of 40 degrees to radians:

1. **Put the 40 in place of the θ° in the proportion.**

   $$\frac{40}{180} = \frac{\theta^R}{\pi}$$

2. **Reduce the fraction on the left.**

   $$\frac{{}^{2}\cancel{40}}{{}^{9}\cancel{180}} = \frac{2}{9} = \frac{\theta^R}{\pi}$$

3. **Multiply each side of the proportion by π.**

   $$\pi \cdot \frac{2}{9} = \frac{\theta^R}{\pi} \cdot \pi$$

4. **Simplify the work.**

   $$\pi \cdot \frac{2}{9} = \frac{\theta^R}{\cancel{\pi}} \cdot \cancel{\pi}$$

   $$\frac{2\pi}{9} = \theta^R$$

This example shows that 40 degrees is equivalent to $\frac{2\pi}{9}$ radians. You leave the radian measure as a fraction reduced to lowest terms.

Check out another example: Change a measure of –36 degrees to radians.

1. **Put the –36 in place of the $\theta°$ in the proportion.**

$$\frac{-36}{180} = \frac{\theta^R}{\pi}$$

2. **Reduce the fraction on the left.**

$$\frac{-{}^1\cancel{36}}{{}^5\cancel{180}} = -\frac{1}{5} = \frac{\theta^R}{\pi}$$

3. **Multiply each side of the proportion by π.**

$$\pi \cdot \left(-\frac{1}{5}\right) = \frac{\theta^R}{\pi} \cdot \pi$$

4. **Simplify the work.**

$$\pi \cdot \left(-\frac{1}{5}\right) = \frac{\theta^R}{\cancel{\pi}} \cdot \cancel{\pi}$$

$$-\frac{\pi}{5} = \theta^R$$

So you see, –36 degrees is equivalent to $-\frac{\pi}{5}$ radians. Having a negative angle is fine (see Chapter 4 for more on negative angles). You leave the expression as a fraction; don't change it to a decimal form.

Changing radians to degrees

You use the same basic proportion to change radians to degrees as you do for changing degrees to radians.

For example, to change $\frac{\pi}{12}$ radians to a degree measure:

1. **Put the radian measure in place of the θ^R in the proportion.**

$$\frac{\theta°}{180} = \frac{\frac{\pi}{12}}{\pi}$$

2. Simplify the complex fraction on the right by multiplying the numerator by the reciprocal of the denominator.

$$\frac{\theta°}{180} = \frac{\cancel{\pi}}{12} \cdot \frac{1}{\cancel{\pi}}$$

$$\frac{\theta°}{180} = \frac{1}{12}$$

3. Multiply each side of the proportion by 180.

$$180 \cdot \frac{\theta°}{180} = \frac{1}{12} \cdot 180$$

4. Reduce and simplify the fraction on the right.

$$\cancel{180} \cdot \frac{\theta°}{\cancel{180}} = \frac{1}{{}^1\cancel{12}} \cdot \cancel{180}^{\,15}$$

$$\theta° = 15$$

So, $\frac{\pi}{12}$ radians is equivalent to 15 degrees.

Here's another example: Change 1.309 radians to degrees.

I changed this radian measure to a decimal by multiplying through by a decimal equivalent of π, which is approximately 3.1416. You use this same decimal equivalent to solve the problem.

Just a minute

A full circle contains 360 degrees. If you want just a part of a degree — and a degree is already pretty small — you can say you have $\frac{1}{2}$ of a degree or 0.5 of a degree, or you can use another division. You can divide one degree into 60 *minutes,* and you can divide each minute into 60 *seconds.* So, mathematically, a degree has 3,600 subdivisions — you can break it down into 3,600 seconds. The way you denote the number of minutes and seconds is with one tick mark (') for minutes and two tick marks (") for seconds. So you read the degree measure 15°45′27″ like this: "Fifteen degrees, 45 minutes, and 27 seconds." Ever since the advent of hand-held calculators, people don't use this measure much anymore. The decimal breakdown of a degree is more universally accepted.

1. **Put the radian measure in place of the θ^R in the proportion.**

$$\frac{\theta°}{180} = \frac{1.309}{\pi}$$

2. **Change the π to a decimal approximation. In this case, I used four decimal places.**

$$\frac{\theta°}{180} = \frac{1.309}{3.1416}$$

3. **Multiply each side of the proportion by 180.**

$$180 \cdot \frac{\theta°}{180} = \frac{1.309}{3.1416} \cdot 180$$

4. **Reduce the fractions, and simplify the value on the right.**

$$\cancel{180} \cdot \frac{\theta°}{\cancel{180}} = \frac{1.309}{3.1416} \cdot 180$$

$$\theta° = \frac{235.62}{3.1416} = 75$$

This result came out to be a nice number. Sometimes, however, you have a decimal answer for the degrees. Actually, you get a decimal more than sometimes — you *usually* get one.

Highlighting favorites

The *favorite* or most-used angles are those that are multiples of 15 degrees, such as 30, 45, 60, and 90 degrees. Putting these angles into the proportion for changing degrees to radians gives a nice set of angles in radian measure.

First, look at what happens when you replace $\theta°$ with 30:

$$\frac{30}{180} = \frac{\theta^R}{\pi}$$

$$\frac{1}{6} = \frac{\theta^R}{\pi}$$

$$\pi \cdot \frac{1}{6} = \frac{\theta^R}{\pi} \cdot \pi$$

$$\frac{\pi}{6} = \theta^R$$

An angle of 30 degrees is equivalent to $\frac{\pi}{6}$ radians. You get a simple fraction with a π on the top and a nice, small 6 on the bottom.

You get similar results with the other angles:

$$45° \to \frac{\pi}{4} \qquad\qquad 60° \to \frac{\pi}{3} \qquad\qquad 90° \to \frac{\pi}{2}$$

Radian measures with denominators of 2, 3, 4, and 6 are used most frequently.

Making a Clone of Arc

The biggest advantage of using radians instead of degrees is that a radian is directly tied to a length — the length or distance around a circle, which is called its *circumference*. Using radians is very helpful when doing applications involving the length of an arc of a circle, which is part of its circumference; measuring the sweep of a hand on a clock; and finding distance in navigation problems.

The problems in this section give you a good sampling of situations where using radians is your best bet. Of course, all these problems presume that you can make *accurate* measurements of the variables you can measure. But trigonometry does open the door to solving practical problems that aren't doable any other way.

Taking chunks out of circles

The examples in this section use features of circles. A part of a circle may be an arc, a diameter (not really a physical part, but a measure), a *sector* (a piece of the inside), or the center. The measures usually start out in degrees, and I change them to radians, when necessary, to complete the problem.

Scanning with a radar

Radar scans a circular area that has a radius of 40 miles. In one second, it sweeps an arc of 60 degrees. What *area* does the radar cover in one second? In five seconds? Look at Figure 5-2, which shows a sweep of 60 degrees.

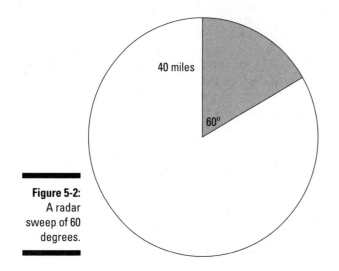

Figure 5-2:
A radar
sweep of 60
degrees.

Here's how you solve this problem:

1. **Find the area of the circle.**

   Use the formula for the area of a circle, $A = \pi r^2$. Putting the 40 in for the radius, r, you get $\pi r^2 = \pi(40)^2 = 1{,}600\pi \approx 5{,}026.548$ square miles.

2. **Divide by the portion of the circle that the sweep covers.**

   The sweep of 60 degrees is only $\frac{1}{6}$ of the entire circle, so you figure the area that the sweep covers by dividing the entire area by 6. The resulting area is $\frac{5{,}026.548}{6} = 837.758$ square miles, which is the area the radar scans in one second. To get the area covered in five seconds, multiply that result by 5 to get 4,188.790 square miles.

The preceding problem works out nicely, because the number of degrees is a convenient value — it's a fraction of the circle. But what if the number doesn't divide evenly into 360? For example, what if the radar sweeps an angle of 76 degrees in one second?

In general, if the angle is given in degrees, then the part of a circle that the angle sweeps is $\frac{\text{angle in degrees}}{360}$. Take the fraction for that part of the circle and multiply it by the area, πr^2. A fancy name for this part of a circle is *sector*.

Keep the following formulas in mind when you're trying to find the area of a sector:

✔ **Using degrees:** Area of sector $= \dfrac{\theta^\circ}{360} \cdot \pi r^2$

✔ **Using radians:** Area of sector $= \dfrac{\theta^R \cdot r^2}{2}$

The second formula comes from the following computation. That's why there's no π in the result:

$$\frac{\theta^R}{2\pi} \cdot \pi r^2 = \frac{\theta^R}{2\pi} \cdot \pi r^2 = \frac{\theta^R \cdot r^2}{2}$$

For example, to find the area of the radar sweep in the preceding example when the radar sweeps 76 degrees per second:

1. **Put 76 in for the θ° and 40 for the radius in the formula for the area of a sector.**

   Area of sweep $= \dfrac{76}{360} \cdot \pi (40)^2$

2. **Multiply and divide to simplify the answer.**

   $= \dfrac{121,600\pi}{360} \approx \dfrac{382,017.667}{360} = 1,061.160$ square miles

To demonstrate this radar-sweep calculation if you're given measurements in radians, find the area of the radar sweep if the sweep is $\frac{\pi}{3}$ radians (which is 60 degrees).

1. **Put $\frac{\pi}{3}$ in for the θ^R and 40 in for the radius.**

   Area of sweep $= \dfrac{\theta^R \cdot r^2}{2} = \dfrac{\frac{\pi}{3}(40)^2}{2}$

2. **Multiply and divide to simplify the answer.**

   $= \dfrac{\frac{\pi}{3} \cdot 1,600}{2} \approx \dfrac{1,675.516}{2} = 837.758$ square miles

Compare this result with the computation for the sweep of 60 degrees, earlier in this section.

Sharing pizza

Some fraternity brothers want to order pizza — and you know how hungry college men can be. The big question is, which has bigger slices of pizza: a 12-inch pizza cut into six slices, or a 15-inch pizza cut into eight slices? Figure 5-3 shows a 12-inch pizza and a 15-inch pizza, both of which are sliced. Can you tell by looking at them which slices are bigger — that is, have more area?

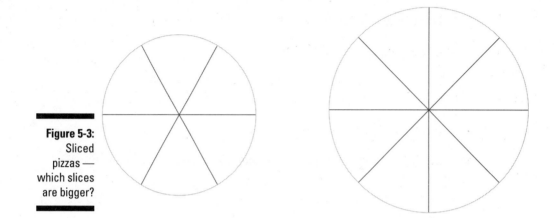

Figure 5-3:
Sliced
pizzas —
which slices
are bigger?

The 12-inch pizza is cut into six pieces. Each piece represents an angle of 60 degrees, which is $\frac{\pi}{3}$ radians, so you find the area of each sector (slice) by using the formula for the area of a sector using radians and putting the 6 in for the radius of the pizza with a 12-inch diameter. The answer is

$$\frac{\theta^R \cdot r^2}{2} = \frac{\frac{\pi}{3} \cdot (6)^2}{2} = \frac{\frac{\pi}{3} \cdot 36}{2} = \frac{12\pi}{2} = 6\pi \text{ square inches.}$$ (I leave the answer with

the multiplier of π just so you can compare the sizes between the two pizzas — they'll both have a multiplier of π in them.)

The 15-inch pizza is cut into eight pieces. Each piece represents an angle of 45 degrees, which is $\frac{\pi}{4}$ radians, so, letting the radius be 7.5 this time, the area of

each sector is $\dfrac{\theta^R \cdot r^2}{2} = \dfrac{\frac{\pi}{4}(7.5)^2}{2} = \dfrac{\frac{\pi}{4} \cdot 56.25}{2} = \dfrac{56.25\pi}{8} = 7.03125\pi$ square inches.

This result doesn't tell you exactly how many square inches are in each slice, but you can see that a slice of this 15-inch pizza has an area of 7.03125π square inches, and a slice of the 12-inch pizza has an area of 6π square inches. The 15-inch pizza has bigger pieces, even though you cut it into more pieces than the 12-inch pizza. And, by the way, the difference is slightly over three square inches per slice.

Sweeping hands

I discuss two scenarios in this section: the minute hand of a clock sweeping across the clock's face and the hand of a rider on a Ferris wheel as it whooshes through the air. These examples use the formula for arc length, which is the distance around part of a circle.

You find the length of an arc of a circle, s, by using the following formula, where the measure of the angle is in radians, and r stands for the radius of the circle: $s = \theta^R \cdot r$.

Riding the minute hand

Suppose a ladybug settled onto the tip of a tower clock's minute hand. The minute hand is 12 feet long. How far does the ladybug travel from 3:00 until 3:20?

1. **Calculate how many degrees the minute hand swings in 20 minutes.**

   Twenty minutes is $\frac{20}{60}$ or $\frac{1}{3}$ of an hour. Translate that fraction into degrees, and you get $\frac{1}{3}$ of 360, or 120 degrees.

2. **Convert degrees to radians.**

   The formula for arc length uses angles in terms of radians, so you first need to change 120 degrees to radians. Using the proportion for changing from degrees to radians and reducing the fraction on the left, $\frac{120}{180} = \frac{\theta^R}{\pi}$ or $\frac{2}{3} = \frac{\theta^R}{\pi}$. Multiply each side of the equation by π. The final result for the angle measure is $\theta^R = \frac{2\pi}{3}$.

3. **Calculate the answer by using the formula for the length of the arc.**

   Enter the angle in radians, and enter 12 (feet), which is the length of the minute hand, for the radius. Your computations should look like $s = \theta^R \cdot r = \frac{2\pi}{3} \cdot 12 = 8\pi$, which is the distance the ladybug traveled, about 25.13 feet.

Riding the Ferris wheel

I don't usually care for Ferris wheels and heights, but on the rare occasions when someone *does* talk me into riding one, I have to face my least favorite part of the whole event: coming down the front of the wheel (I call this part the front of the wheel because you can't see the ground below you). Take a look at Figure 5-4. Imagine that this is the London Eye, a giant Ferris wheel in London, England. The diameter of this wheel is 394 feet. If I'm in a car on the wheel and travel from the top of the wheel, halfway down to the bottom, then how far have I traveled (hyperventilating the whole way)?

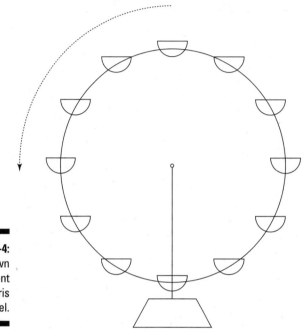

Figure 5-4:
Riding down
the front
of a Ferris
wheel.

A circle with a 394-foot diameter has a 197-foot radius. From the top to the front of the wheel is $\frac{1}{4}$ of the circle, which is 90 degrees. In radians, 90 degrees is $\frac{\pi}{2}$ (see the section "Highlighting favorites," earlier in this chapter). Using the formula for arc length and putting in the radian measure and the radius of 197 feet, the distance is $s = \theta^R \cdot r = \frac{\pi}{2} \cdot 197 = 98.5\pi$, which is about 309.45 feet. Egad!

Going out and about

One of trigonometry's great qualities is that it lets you measure things that you can't get at or, in the case of the racetrack example in this section, things that you don't want to get close to. A circle and its angles have all sorts of applications both on earth and above.

Measuring the distance to the moon

One of the earliest applications of trigonometry was in measuring distances that you couldn't reach, such as distances to planets or the moon or to places on the other side of the world. Consider the following example.

The diameter of the moon is about 2,160 miles. When the moon is full, a person sighting the moon from the earth measures an angle of 0.56 degrees from one side of the moon to the other (see Figure 5-5).

Figure 5-5:
A person on the earth sights the top and bottom of the moon.

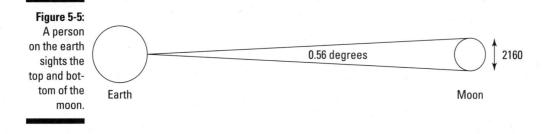

0.56 degrees   2160

Earth   Moon

To figure out how far away the moon is from the earth, consider a circle with the earth at the center and the circumference running right through the center of the moon, along one of the moon's diameters. The moon is so far away that the straight diameter and slight curve of this big circle's circumference are essentially the same measure. The arc that runs through the moon's diameter has an angle of 0.56 degrees and an arc length of 2,160 miles (the diameter). Using the arc-length formula, solve for the radius of the large circle, because the radius is the distance to the moon. To solve for the radius:

1. **First, change 0.56 degrees to radians.**

$$\frac{0.56}{180} = \frac{\theta^R}{\pi} \qquad \theta^R = \frac{0.56\pi}{180} \approx 0.00977$$

2. **Input the numbers into the arc-length formula, $s = \theta^R \cdot r$.**

   Enter 0.00977 radians for the radian measure and 2,160 for the arc length: $2{,}160 = 0.00977 \cdot r$.

3. **Divide each side by 0.00977.**

   The distance to the moon is $r = \dfrac{2{,}160}{0.00977} \approx 221{,}085$ miles.

Racing around the track

A race car is going around a circular track. A photographer standing at the center of the track takes a picture, turns 80 degrees, and then takes another picture 10 seconds later. If the track has a diameter of $\frac{1}{2}$ mile, how fast is the race car going? Figure 5-6 shows the photographer in the middle and the car in the two different positions.

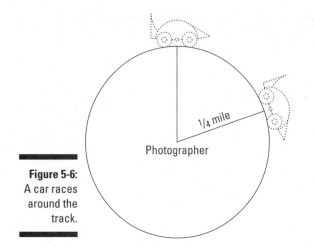

Figure 5-6:
A car races
around the
track.

How fast is the car going? Where does the problem make any mention of speed? Actually, in this situation, the car travels partway around the track in 10 seconds. By computing the arc length, you can determine how fast the car is traveling.

A formula that you'll find mighty helpful is the one that says distance equals rate multiplied by time, where rate is miles per hour (or feet per second or some such measure), and time is the same measure as in the rate: $d = r \cdot t$.

1. **First, change the 80 degrees to radians.**

   You end up with $\frac{4\pi}{9}$ radians.

2. **Input the numbers in the arc-length formula.**

   Putting in the radian measure and the radius of the track, $\frac{1}{4}$ mile, you

   get Arc Length $= \frac{4\pi}{9} \cdot \frac{1}{4} = \frac{\pi}{9} \approx 0.349$ mile, which is the distance the car

   traveled in 10 seconds.

3. **Multiply this result by 6 (since 10 seconds is $\frac{1}{6}$ of a minute) to get miles traveled in a minute.**

   This calculation gives you 2.094 miles per minute.

4. **Then multiply that number by 60 to get miles traveled in one hour.**

   This calculation gives you 125.64 miles. So the car is traveling about 126 mph.

Chapter 6

Getting It Right with Triangles

· ·

· ·

*T*riangles are classified in many ways. One way of distinguishing one triangle from another is to use angle measurements. Because a 90-degree angle is called a *right angle,* you use the same terminology to describe a triangle with a right angle in it. This type of triangle is called a *right triangle.* And that's all right.

The measures of the sides of right triangles are used to determine the values of the trig functions. And those trig functions (along with the right triangles) are really handy when it comes to solving problems such as "Just how high is that tree?" The special properties of a right triangle — some credited with Pythagoras — make them very useful in trigonometry and other math areas.

Sizing Up Right Triangles

If you're looking at their angles, triangles can be right, acute, or obtuse. Right triangles have been of great interest for centuries. They're the basis for applications in navigation, astronomy, surveying, and military engineering.

What's so right about them?

A *right triangle* has a right angle in it. But it can only have one right angle, because the total number of degrees in a triangle is 180. If it had two right angles, then those two angles would take up all 180 degrees; no degrees would be left for a third angle. So in a right triangle, the other two angles share the remaining 90 degrees.

Right triangles can come in all sorts of shapes, but they all have that corner, where the right angle sits. In Figure 6-1, you see that in all the triangles, the right angle has the two sides that are perpendicular to one another. The other two angles are *acute* angles (meaning they're less than 90 degrees).

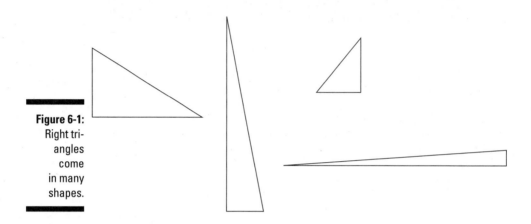

Figure 6-1:
Right tri-
angles
come
in many
shapes.

The anatomy of a right triangle

Right triangles are a familiar sight — not just in geometry classes. Carpenters have tools for measurement that are right triangles. Architects who design by hand (rather than on a computer) draw with right-triangle templates. Even though the focus in a right triangle is the right angle, a right triangle actually has six different parts: three angles and three sides. Now, this fact is true of any triangle, but right triangles have special names for these parts. Having special names is necessary because so many properties, theorems, and applications using right triangles are out there, and the names make talking about and explaining the triangles more clear.

Figure 6-2 shows a typical right triangle labeled with capital letters and lowercase letters. Since the time of Leonard Euler, the famous Swiss mathematician, this type of labeling has been the tradition. You use lowercase letters to mark the sides of the triangle and capital letters to mark the *vertex* (angle) opposite the side with the corresponding lowercase letter.

The little square at the vertex C shows that the two sides meeting there are perpendicular at that vertex — that's where the right angle is. The side c, opposite the right angle, is called the *hypotenuse*. The other two sides, a and b, are called the *legs*. The hypotenuse is always the longest side, because it's opposite the largest angle.

Squaring the corners

Over 30 years ago, my spouse and I selected house plans that we liked and started all the processes needed to build our new home. We didn't have all that much money, so we did as much as we could by ourselves. After the lot was cleared of bushes and weeds, we went there with pegs and string to lay out the foundation. The backhoe was due the next day to dig the hole for the basement. With blueprints in hand, we had all the measurements — how long each side of the house was to be. A tape measure gave us accurate measures for lengths, but what about the right angles? A school protractor isn't big enough to make those long sides exactly perpendicular. We used a method called *squaring the corners.* We knew that some nice measures for the three sides of a right triangle are 3-4-5 or 6-8-10. At a corner in question, we would pick a point on one string that was 4 feet from the peg and a point on the other string that was 3 feet from the peg. Then we'd measure diagonally from each of those points. If the diagonal measure didn't come out to be exactly 5 feet, then we didn't have a right angle — the corner wasn't square. It took a lot of peg moving, but we got the foundation laid out accurately. The house is still standing — in fact, we still live in it — so I guess we did a good job!

In Figure 6-2, the angle at vertex C is the right angle, and the other two angles, A and B, are acute angles. If the measures of the angles at A and B are the same, then they're each 45 degrees, and the triangle is *isosceles.* If that's the case, then the lengths of the sides a and b are the same also.

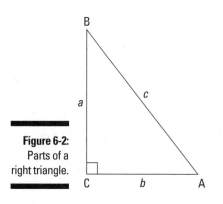

Figure 6-2:
Parts of a
right triangle.

If the angle at vertex A is bigger than that at vertex B, then side a is longer than side b. The measures of a triangle's angles have a direct relationship to the lengths of the sides opposite them.

Pythagoras Schmythagoras: Demystifying the Pythagorean Theorem

Pythagoras was a Greek mathematician who lived around 570 B.C. Even with the relatively primitive tools at his disposal, he was able to discover and formulate a *theorem,* or rule, that became one of the most well known in all of mathematics: the Pythagorean theorem.

The *Pythagorean theorem* says that if a, b, and c are the sides of a right triangle, as shown in Figure 6-3, and if c is the side opposite the right angle, then their lengths have the following property: $a^2 + b^2 = c^2$.

Figure 6-3: Traditional labels for the sides of a right triangle.

Have you heard the one about the squaw on the hippopotamus?

The first time I heard this joke was on a Chicago radio station in the 1960s. The disc jockey was Howard Miller — as many Chicagoans will agree, a man before his time. Who else could get away with telling a math joke on morning radio? The joke goes as follows: There was an Indian tribe with the usual hierarchy of braves, chiefs, and squaws. Two of the braves had squaws who were expecting papooses. The chief's squaw was also expecting a papoose. The squaws of the braves sat working on buffalo hides, grinding corn and doing what squaws do. But the squaw of the chief sat on a hippopotamus hide to do her daily chores. The happy days arrived; the papooses were born. The squaws of the two braves each had sons. And the squaw of the chief had *twin* sons. The moral of this story is "The squaw on the hippopotamus is equal to the sons of the squaws on the other two hides." Yes, I'm sure you're groaning.

You can work backward with the Pythagorean theorem, too. For example, if you don't know what type of triangle you have, and the sides of a triangle are 3, 4, and 5 units in length, then the triangle must be a right triangle, because

$$3^2 + 4^2 = 5^2$$
$$9 + 16 = 25$$

The sum of the squares of the two shorter sides, which are called the *legs*, is equal to the square of the longest side, which is called the *hypotenuse*.

Hitting a Pythagorean triple

A *Pythagorean triple* is a list of three numbers that works in the Pythagorean theorem — the square of the largest number is equal to the sum of the squares of the two smaller numbers. The multiple of any Pythagorean triple (multiply each of the numbers in the triple by the same number) is also a Pythagorean triple. They seem to reinvent themselves.

TIP

Familiarizing yourself with the more frequently used Pythagorean triples is very helpful. If you recognize that you have a triple, then working with applications is much easier.

Table 6-1 shows some of the most common Pythagorean triples and some of their multiples.

Table 6-1	Common Pythagorean Triples		
Triple	*Triple* × *2*	*Triple* × *3*	*Triple* × *4*
3-4-5	6-8-10	9-12-15	12-16-20
5-12-13	10-24-26	15-36-39	20-48-52
7-24-25	14-48-50	21-72-75	28-96-100
9-40-41	18-80-82	27-120-123	36-160-164
11-60-61	22-120-122	33-180-183	44-240-244

Here's how to check out a triple and its multiple by using the Pythagorean theorem. Try out the triple 9-40-41:

1. **Replace *a*, *b*, and *c* with 9, 40, and 41, respectively.**

$$9^2 + 40^2 = 41^2$$
$$81 + 1{,}600 = 1{,}681$$

2. **Then replace *a*, *b*, and *c* with the 9-40-41 triple multiplied by 3 (which is 27-120-123).**

$$27^2 + 120^2 = 123^2$$
$$729 + 14{,}400 = 15{,}129$$

Solving for a missing length

One of the nice qualities of right triangles is the fact that you can find the length of one side if you know the lengths of the other two sides. You don't have this luxury with just any triangle, so count your blessings now.

Practicing on triangles

The Pythagorean theorem states that $a^2 + b^2 = c^2$ in a right triangle where c is the longest side. You can use this equation to figure out the length of one side if you have the lengths of the other two. Figure 6-4 shows two right triangles that are each missing one side's measure.

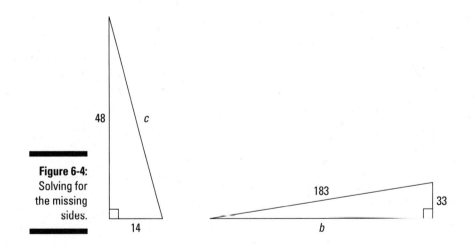

Figure 6-4:
Solving for
the missing
sides.

In the left triangle in Figure 6-4, the measure of the hypotenuse is missing. Use the Pythagorean theorem to solve for the missing length.

1. **Replace the variables in the theorem with the values of the known sides.**

   $48^2 + 14^2 = c^2$

2. **Square the measures and add them together.**

   $$2,304 + 196 = c^2$$
   $$2,500 = c^2$$
   $$\sqrt{2,500} = \sqrt{c^2}$$
   $$50 = c$$

The length of the missing side, c, which is the hypotenuse, is 50.

The triangle on the right in Figure 6-4 is missing the bottom length, but you do have the length of the hypotenuse. It doesn't matter whether you call the missing length a or b.

1. **Replace the variables in the theorem with the values of the known sides.**

   $33^2 + b^2 = 183^2$

2. **Square the measures, and subtract 1,089 from each side.**

   $$1,089 + b^2 = 33,489$$
   $$b^2 = 32,400$$

3. **Find the square root of each side.**

   $$\sqrt{b^2} = \sqrt{32,400}$$
   $$b = 180$$

The length of the missing side is 180 units. That's not much shorter than the hypotenuse, but it still shows that the hypotenuse has the longest measure.

Finding the distance across a pond

Trigonometry is very handy for finding distances that you can't reach to measure. Imagine that you want to string a cable diagonally across a pond (so you can attach a bunch of fishing line and hooks). The diagonal distance is the hypotenuse of a right triangle. You can measure the other two sides along the shore. Figure 6-5 shows the pond and the imaginary right triangle you use to figure out how long your cable needs to be.

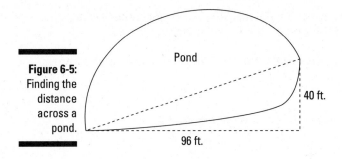

Figure 6-5:
Finding the
distance
across a
pond.

Pond

40 ft.

96 ft.

The two sides of the triangle that you can measure, the height and the width of the pond, are 40 feet and 96 feet. These are the two legs of a right triangle. Use the Pythagorean theorem to solve for the hypotenuse, which is the diagonal distance across the pond.

1. **Replace the variables in the theorem with the values of the known sides.**

   $40^2 + 96^2 = c^2$

2. **Square the measures, and add them together.**

   $$1,600 + 9,216 = c^2$$
   $$10,816 = c^2$$

3. **Find the square root of the sum.**

   $$\sqrt{10,816} = \sqrt{c^2}$$
   $$104 = c$$

The diagonal across the pond is 104 feet. String up your cable, and go fishing!

In a League of Their Own: Special Right Triangles

Right triangles are handy little suckers. The relationship between the lengths of the sides helps you measure lengths that you can't reach. And just when you think that math can't get any better, along come two triangles that are the cat's meow (see Figure 6-6). One of them is an isosceles right triangle. The two legs are the same, and the hypotenuse is always a multiple of their length. The other special right triangle has one side half as long as the other.

These two triangles are very useful, because the angle measures in them are some of the most popular, and the side measures are used in trig functions.

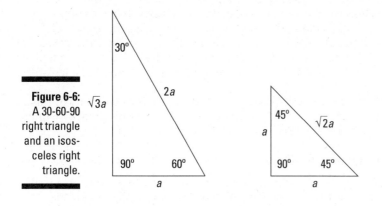

Figure 6-6: A 30-60-90 right triangle and an isosceles right triangle.

30-60-90 right triangles

A 30-60-90 right triangle has angles measuring just what the name says. The two acute, complementary angles are 30 and 60 degrees. These triangles are great to work with, because the angle measures, all being multiples of 30, have a pattern, and so do the measures of the sides. Oh, yes, the Pythagorean theorem still holds — you have that relationship between the squares of the sides. But a, b, and c are related in another way, too. In a 30-60-90 right triangle, if a is the shortest side, then the hypotenuse, the longest side, measures twice that, or $2a$. You can use $2a$ instead of the letter c. And the middle length is $\sqrt{3}a$, or about 1.7 times as long as the shortest side; this number replaces the letter b. The particularly nice part about this triangle is that you can write all three sides in terms of one variable, a. Look at how these lengths fit into the Pythagorean theorem:

$$a^2 + \left(\sqrt{3}a\right)^2 = \left(2a\right)^2$$
$$a^2 + 3a^2 = 4a^2$$

Here's a sample problem you can solve by taking advantage of the special relationships within a 30-60-90 right triangle: If the hypotenuse of a 30-60-90 right triangle is 8 units long, then how long are the other two sides?

1. **Find the length of the shorter leg.**

   The hypotenuse is twice as long as the shorter leg, a. So $8 = 2a$. Divide by 2, and you get $a = 4$.

2. **Find the length of the longer leg.**

   The longer leg is $\sqrt{3}a$, so multiply $\sqrt{3}$ times 4 to get $4\sqrt{3}$, or about 6.9 units.

Isosceles right triangles

The other special right triangle is the *isosceles right triangle,* or the 45-45-90 right triangle. The two acute angles are equal, making the two legs opposite them equal, too. What's more, the lengths of those two legs have a special relationship with the hypotenuse (in addition to the one in the Pythagorean theorem, of course). In an isosceles right triangle, if the legs are each a units in length, then the hypotenuse is $\sqrt{2}a$, or about 1.4 times as long as a leg.

Now that you know how isosceles right triangles work, try your hand at this sample problem: If an isosceles right triangle has a hypotenuse that's 16 units long, then how long are the legs?

1. **Create an equation to solve.**

   The hypotenuse is $\sqrt{2}a$, where a is the length of the legs. You know that the hypotenuse is 16, so you can solve the equation $\sqrt{2}a = 16$ for the length of a.

2. **Solve for a.**

   Divide each side by the radical to get

   $$\frac{\sqrt{2}a}{\sqrt{2}} = \frac{16}{\sqrt{2}}$$

   $$a = \frac{16}{\sqrt{2}} = \frac{16}{\sqrt{2}} \cdot \frac{\sqrt{2}}{\sqrt{2}} = \frac{16\sqrt{2}}{2} = 8\sqrt{2}$$

   Each leg is about 11.3 units.

Part II
Trigonometric Functions

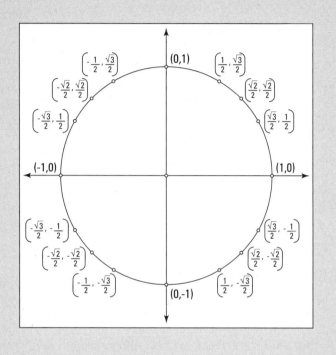

Find out more about finding values for trig functions in an article at www.dummies.com/extras/trigonometry.

In this part...

✔ Define the basic trig functions using the lengths of the sides of a right triangle.

✔ Determine the relationships between the trig cofunctions and their shared sides.

✔ Extend your scope to angles greater than 90 degrees using the unit circle.

✔ Investigate the ins and outs of the domains and ranges of the six trig functions.

✔ Use reference angles to compute trig functions.

Chapter 7

Doing Right by Trig Functions

• •

• •

*B*y taking the lengths of the sides of right triangles or the chords of circles and creating ratios with those numbers and variables, our ancestors marked the birth of trigonometric functions. These functions are of infinite value, because they allow you to use the stars to navigate and to build bridges that won't fall. If you're not into navigating a boat or engineering, then you can use the trig functions at home to plan that new addition. They're a staple for students going into calculus.

The six trig functions require one thing of you — inputting an *angle* measure — and then they output a number. These outputs can be any real numbers, from infinitely small to infinitely large and everything in between. The results you get depend on which function you use. Although some of the early-day computations were rather tedious, today's hand-held calculators make everything much easier.

SohCahToa to the Rescue: How Trig Functions Work

What or who is *SohCahToa* — an Italian pasta dish, a Native American princess, or some new miracle drug, perhaps? Actually, none of the above. Some clever math teacher made up this word in order to help students remember the trig ratios. The word then got around. Before I explain *SohCahToa,* you need to know what the letters stand for. You can see the letters *S, C,* and *T* for the trig functions sine, cosine, and tangent. The lowercase letters represent their ratios in a right triangle.

The name game: A right triangle's three sides

A right triangle has two shorter sides, or legs, and the longest side, opposite the right angle, which is always called the *hypotenuse.* The two shorter sides have some other special names, too, based on which acute angle of the triangle you happen to be working with at a particular time.

In reference to acute angle θ (see Figure 7-1), the leg on the other side of the triangle from θ is called the *opposite side.* That opposite side is never along one of the rays making up the angle. The other leg in the right triangle is then called the *adjacent side. Adjacent* means "next to," and in the case of right triangles, the adjacent side helps form the acute angle along with the hypotenuse because it lies along one of the angle's rays.

Figure 7-1:
The acute
angle θ
determines
the names
of the right
sides.

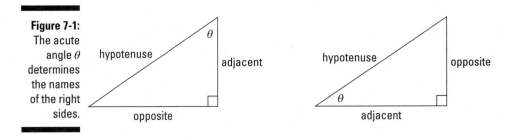

The six ratios: Relating the three sides

Each of the three sides of a right triangle — hypotenuse, opposite, and adjacent — has a respective length or measure. And those three lengths or measures form six different ratios. Check out Figure 7-2, which has sides of lengths 3, 4, and 5.

Figure 7-2: A
right
triangle with
sides of
lengths 3, 4,
and 5.

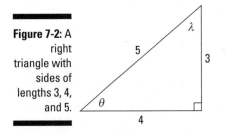

How far to the moon?

Hipparchus lived from about 190 to 120 B.C. Also known as Hipparchus of Nicaea, he did his astrological observations from Rhodes between 146 and 127 B.C. He was the first astronomer to compile a catalog of 850 stars, well before telescopes were available, and his computations were remarkably accurate. He obtained measurements of the length of the year and the distance to the moon. To measure the distance to the moon, Hipparchus and a colleague each observed a solar eclipse — a total eclipse at Syene and a partial eclipse at Alexandria, where four-fifths of the sun was blocked. Using angles and the distance between the two cities, thus creating an imaginary triangle with the lines of sight from those two cities to the moon, he was able to make his calculations. The trig ratios he used were on a big scale, but it doesn't matter how large the triangle is, because the trig functions for the angles don't change with the triangle's size. Hipparchus is commemorated by having a moon crater, a Mars crater, and an asteroid named after him.

The six different ratios that you can form with the numbers 3, 4, and 5 are $\frac{3}{4}, \frac{4}{5}$, $\frac{3}{5}, \frac{4}{3}, \frac{5}{4}$, and $\frac{5}{3}$. These six fractions are all that you can make by using the three lengths of the sides. The ratios are special because they represent all the possible output values of the trig functions for the acute angles in that triangle. And even better, you can figure out the value of an unknown angle in a right triangle just by creating one of these ratios and figuring out which angle has that trig function.

The sine function: Opposite over hypotenuse

When you're using right triangles to define trig functions, the trig function *sine,* abbreviated *sin,* has input values that are angle measures and output values that you obtain from the ratio $\frac{\text{opposite}}{\text{hypotenuse}}$. Figure 7-2 (in the preceding section) shows two different acute angles, and each has a different value for the function sine. The two values are $\sin\theta = \frac{3}{5}$ and $\sin\lambda = \frac{4}{5}$.

The sine is always the measure of the opposite side divided by the measure of the hypotenuse. Because the hypotenuse is always the longest side, the number on the bottom of the ratio will always be larger than that on the top. For this reason, the output of the sine function will always be a proper fraction — it'll never be a number equal to or greater than 1 unless the opposite side is equal in length to the hypotenuse (which only happens when your triangle is a single segment or you're working with circles — see Chapter 8).

Even if you don't know both lengths required for the sine function, you can calculate the sine if you know any two of the three lengths of a triangle's sides. For example, to find the sine of angle α in a right triangle whose hypotenuse is 10 inches long and adjacent side is 8 inches long:

1. **Find the length of the side opposite α.**

   Use the Pythagorean theorem, $a^2 + b^2 = c^2$ (see Chapter 6), letting a be 8 and c be 10. When you input the numbers and solve for b, you get

   $$8^2 + b^2 = 10^2$$
   $$64 + b^2 = 100$$
   $$b^2 = 36$$
   $$b = 6$$

   So, the opposite side is 6 inches long.

2. **Use the ratio for sine, opposite over hypotenuse.**

   $$\sin \alpha = \frac{\text{opposite}}{\text{hypotenuse}} = \frac{6}{10} = \frac{3}{5}$$

The cosine function: Adjacent over hypotenuse

The trig function *cosine,* abbreviated *cos,* works by forming this ratio: $\frac{\text{adjacent}}{\text{hypotenuse}}$. Take a look back at Figure 7-2, and you see that the cosines of the two angles are $\cos \theta = \frac{4}{5}$ and $\cos \lambda = \frac{3}{5}$. The situation with the ratios is the same as with the sine function — the values are going to be less than or equal to 1 (the latter only when your triangle is a single segment or when dealing with circles), never greater than 1, because the hypotenuse is the denominator.

TIP

The two ratios for the cosine are the same as those for the sine — except the angles are reversed. This property is true of the sines and cosines of *complementary angles* in a right triangle (meaning those angles that add up to 90 degrees).

If θ and λ are the two acute angles of a right triangle, then $\sin \theta = \cos \lambda$ and $\cos \theta = \sin \lambda$.

Now for an example. To find the cosine of angle β in a right triangle if the two legs are each $\sqrt{6}$ feet in length:

1. **Find the length of the hypotenuse.**

   Using the Pythagorean theorem, $a^2 + b^2 = c^2$ (see Chapter 6), and replacing both a and b with the given measure, solve for c.

   $$\left(\sqrt{6}\right)^2 + \left(\sqrt{6}\right)^2 = c^2$$
   $$6 + 6 = c^2$$
   $$12 = c^2$$
   $$\sqrt{12} = c$$
   $$2\sqrt{3} =$$

   The hypotenuse is $2\sqrt{3}$ feet long.

2. **Use the ratio for cosine, adjacent over hypotenuse, to find the answer.**

   $$\cos\beta = \frac{\text{adjacent}}{\text{hypotenuse}} = \frac{\sqrt{6}}{2\sqrt{3}} = \frac{\sqrt{2}}{2}$$

The tangent function: Opposite over adjacent

The third trig function, *tangent,* is abbreviated *tan.* This function uses just the measures of the two legs and doesn't use the hypotenuse at all. The tangent is described with this ratio: $\frac{\text{opposite}}{\text{adjacent}}$. No restriction or rule on the respective sizes of these sides exists — the opposite side can be larger, or the adjacent side can be larger. So, the tangent ratio produces numbers that are very large, very small, and everything in between. If you hike on back to Figure 7-2, you see that the tangents are $\tan\theta = \frac{3}{4}$ and $\tan\lambda = \frac{4}{3}$. And in case you're wondering whether the two tangents of the acute angles are always *reciprocals* (flips) of one another, the answer is yes. The trig identities in Chapter 11 explain this phenomenon.

The following example shows you how to find the values of the tangent for each of the acute angles in a right triangle where the hypotenuse is 25 inches and one leg is 7 inches.

1. **Find the measure of the missing leg.**

   Using the Pythagorean theorem, $a^2 + b^2 = c^2$ (see Chapter 6), putting in 7 for a and 25 for c, and solving for the missing value, b, you find that the unknown length is 24 inches:

   $$7^2 + b^2 = 25^2$$
   $$b^2 = 25^2 - 7^2$$
   $$= 625 - 49$$
   $$= 576$$
   $$b = 24$$

2. **Select names for the acute angles in order to determine the *opposite* and *adjacent* designations.**

   The easiest way to do this is to draw a picture and label it — take a look at Figure 7-3.

   The two acute angles are named with the Greek letters θ and λ. The side opposite θ measures 7 inches, and the side adjacent to it measures 24 inches. For angle λ, the opposite side measures 24 inches, and the adjacent side measures 7 inches.

Figure 7-3:
Labeling a right triangle and naming the acute angles.

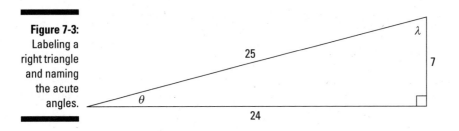

3. **Form the two tangent ratios by using the values 7, 24, and 25.**

   $$\tan\theta = \frac{\text{opposite}}{\text{adjacent}} = \frac{7}{24}$$
   $$\tan\lambda = \frac{\text{opposite}}{\text{adjacent}} = \frac{24}{7}$$

All together, now: Using one function to solve for another

Sometimes you have to solve for a trig function in terms of another function.

In the following example, the cosine of angle λ is $\frac{12}{13}$. What are the values of the sine and tangent of λ?

1. **Identify the sides given by the cosine function.**

   The cosine ratio is $\dfrac{\text{adjacent}}{\text{hypotenuse}}$. Using the given ratio, the adjacent side measures 12 units, and the hypotenuse measures 13 units.

2. **Find the measures of the missing side.**

   Using the Pythagorean theorem, you find that the missing side (the opposite side) measures 5 units.

   $$12^2 + b^2 = 13^2$$
   $$b^2 = 13^2 - 12^2$$
   $$= 169 - 144$$
   $$= 25$$
   $$b = 5$$

3. **Determine the values of the sine and tangent.**

   The sine is $\dfrac{\text{opposite}}{\text{hypotenuse}}$ and the tangent is $\dfrac{\text{opposite}}{\text{adjacent}}$, so $\sin\lambda = \frac{5}{13}$ and $\tan\lambda = \frac{5}{12}$.

Similar right triangles within a right triangle

Take a right triangle and draw an altitude to the hypotenuse. What do you get? You are the proud recipient of three similar triangles — three triangles in decreasing sizes, all with the same three angle measures. Take a look at Figure 7-4 for an example of this situation. The acute angles are labeled α and β to help you find the similarity features.

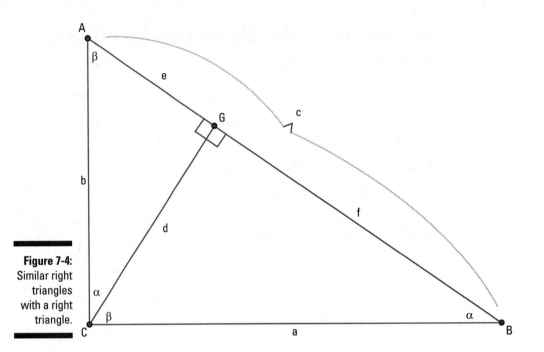

Figure 7-4:
Similar right
triangles
with a right
triangle.

The angles α and β are complementary and appear in each of the three triangles. Triangle ABC is similar to triangle ACG, and both are similar to triangle CBG. I was careful to give these names with the first letter at the vertex with angle α, the second letter at the vertex with angle β, and the last letter at the right angle.

Carefully naming the triangles allows you to write the trig functions of the three different triangles in equations that compare their values. For example, because $\triangle ABC \sim \triangle ACG \sim \triangle CBG$, you can write equations involving ratios of the corresponding sides in the triangles. If you write the ratios of the longer legs divided by the shorter legs, you have:

$$\frac{a}{b} = \frac{d}{e} = \frac{f}{d}$$

Now look at the tangents of the angle β in each triangle. In $\triangle ABC$, $\tan\beta = \frac{a}{b}$. In $\triangle ACG$, $\tan\beta = \frac{d}{e}$. In $\triangle CBG$, $\tan\beta = \frac{f}{d}$.

These and other relationships allow for several equations such as the *means* and *extremes* rules:

- $d^2 = e \cdot f$: The square of the altitude is equal to the product of the two parts of the hypotenuse.

- $b^2 = e \cdot c$: The square of a leg is equal to the product of the part of the hypotenuse adjacent to the leg and the hypotenuse.

Socking the rules away: The legend of SohCahToa

And now, for the fun part: the legend of an Indian chief named SohCahToa (read that: *soak*-uh-*toe*-uh). Many years ago, a tribe of American Indians lived along the Illinois River, where they hunted and fished and did what was necessary to live in peace. One young brave was trying to learn to use his bow and arrows effectively, and he was having all sorts of trouble. Out of frustration, he kicked what he thought was something soft, but it was a rock. His toe turned blue and throbbed all day and night. He tried wrapping it, rubbing it, and ignoring it, but nothing gave him any relief. His mother, a wise squaw, finally had enough of his complaining and said, "Go down to the river, now, and Soh Cah Toa!" The young brave went to the river, put his toe in the cool water, and got relief. He never did get proficient with the bow and arrow, and he kept kicking things in frustration. Pretty soon, he was known as SohCahToa.

Sure, this story is pretty lame, but you'll find it very useful when trying to remember the ratios for the three basic trig functions:

- ✔ **Soh** stands for Sine Opposite Hypotenuse.
- ✔ **Cah** stands for Cosine Adjacent Hypotenuse.
- ✔ **Toa** stands for Tangent Opposite Adjacent.

People who studied trigonometry in the past may not remember too many details about it, but one thing they *do* remember is SohCahToa, if they've heard the story (or something similar).

Taking It a Step Further: Reciprocal Functions

The three most basic trig functions (sine, cosine, and tangent) use the three sides of a triangle, taking two at a time and making ratios/fractions of them. But three more trig functions exist, and these are called the *reciprocal functions* because they use the *reciprocals,* or flips, of the original three functions. If these last three functions are just reciprocals of the first three, why are they even necessary? You could probably live without them, but you'd eventually miss them when doing calculations and solving equations. Having an unknown variable in the numerator of a fraction when solving an equation is just nicer and more convenient than having one in the denominator, and these reciprocal functions make that situation possible. It's nothing more than convenience, but I'm all for such luxury.

The cosecant function: Sine flipped upside down

The *cosecant* function, abbreviated *csc*, is the reciprocal of the sine function and thus uses this ratio: $\dfrac{\text{hypotenuse}}{\text{opposite}}$. The hypotenuse of a right triangle is always the longest side, so the numerator of this fraction is always larger than the denominator. As a result, the cosecant function always produces values bigger than 1.

You can use the values in Figure 7-5 to determine the cosecants of the two acute angles: $\csc\theta = \dfrac{13}{12}$ and $\csc\lambda = \dfrac{13}{5}$.

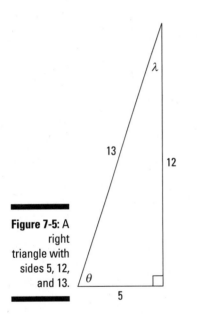

Figure 7-5: A right triangle with sides 5, 12, and 13.

Suppose someone asks you to find the cosecant of angle α if you know that the hypotenuse is 1 unit long and that the right triangle is isosceles. Remember that an isosceles triangle has two congruent sides (flip back to Chapter 6 for a refresher). These two sides have to be the two legs, because the hypotenuse has to have the longest side. So, to find the cosecant:

1. **Find the lengths of the two legs.**

   The Pythagorean theorem says that $a^2 + b^2 = c^2$, but because two sides are congruent, you can take out one variable and write the equation as $a^2 + a^2 = c^2$. Put in 1 for c and solve for a.

$$2a^2 = 1$$

$$a^2 = \frac{1}{2}$$

$$a = \sqrt{\frac{1}{2}}$$

$$= \frac{1}{\sqrt{2}}$$

The legs are both $\frac{1}{\sqrt{2}}$ units long. You can leave the radical in the denominator and not worry about rationalizing, because you're going to input the whole thing into the cosecant ratio, anyway, and things can change.

2. **Use the length of the opposite side in the ratio for cosecant.**

$$\csc \alpha = \frac{\text{hypotenuse}}{\text{opposite}} = \frac{1}{\frac{1}{\sqrt{2}}} = \sqrt{2}$$

The secant function: Cosine on its head

The *secant* function, abbreviated *sec,* is the reciprocal of the cosine. So, its ratio is $\frac{\text{hypotenuse}}{\text{adjacent}}$. Just as with the cosecant, the ratio of the sides is greater than 1. Using the triangle in Figure 7-5, the two secants are $\sec\theta = \frac{13}{5}$ and $\sec\lambda = \frac{13}{12}$.

The cotangent function: Tangent, tails side up

The last reciprocal function is the *cotangent,* abbreviated *cot.* This function is the reciprocal of the tangent (hence, the *co-*). The ratio of the sides for the cotangent is $\frac{\text{adjacent}}{\text{opposite}}$. So, if you look back at Figure 7-5, you see that the two cotangents are $\cot\theta = \frac{5}{12}$ and $\cot\lambda = \frac{12}{5}$.

The ratio for the cotangent is just that ratio, not necessarily the lengths of the sides. The fraction made by the lengths might've been reduced by dividing numerator and denominator by the same number.

Sometimes you know the value of the cotangent along with other information and have to solve for one or both of the sides. Try this example: What are the lengths of the legs of a right triangle if $\cot\alpha = \frac{11}{60}$ and the hypotenuse is 183 inches long?

1. **Write the adjacent and opposite sides as multiples of the same number, *m*, and put them in the Pythagorean theorem with the hypotenuse.**

   $(11m)^2 + (60m)^2 = 183^2$

2. **Simplify the equation and solve for *m*.**

$$(11m)^2 + (60m)^2 = 183^2$$
$$121m^2 + 3,600m^2 = 33,489$$
$$3,721m^2 = 33,489$$
$$m^2 = \frac{33,489}{3721}$$
$$= 9$$
$$m = 3$$

3. **Use the value of *m* to find the lengths of the two legs.**

   Because you know that $m = 3$, you know that the adjacent side is $11m = 11(3) = 33$, and the opposite side is $60m = 60(3) = 180$. The three sides of the right triangle are 33, 180, and 183. You can double-check your results by plugging these three numbers into the Pythagorean theorem and making sure the theorem holds true.

Angling In on Your Favorites

You may have a favorite television show, dessert, or color. Usually, however, a favorite *angle* isn't near the top of anyone's list. But a favorite angle isn't really out of line in the scheme of things. My favorite angle is a 30-degree angle — there's just something so acute about it.

Identifying the most popular angles

The most common or popular angles are those with measures that are multiples of 15 degrees. Topping the list are 30, 60, and 90 degrees. Another favorite is 45 degrees. The reason that these angles are favorites is because they all divide 360 degrees evenly. These exact divisions result in nicer-than-usual values for the different trig functions of the angles.

One way to capitalize even more on the four main angles — 30, 45, 60, and 90 — is to look at their multiples that go up to 360 degrees. The trig functions of the first four basic angles and the trig functions of their multiples are related (see Chapter 8). The list of all-time favorites includes multiples of 30 degrees (30, 60, 90, 120, 150, 180, 210, 240, 270, 300, and 330) and some multiples of 15 degrees that are between them: 45, 135, 225, and 315. All these multiples split the four quadrants the first time around. A 0-degree angle is also highly favored. A measure of 0 degrees is technically a multiple of any of these measures, and you need it because it's the starting point.

Determining the exact values of functions

Even though a scientific calculator gives you the values of the trig functions of any angle, not just your favorite angles, the values it shows you for most of those angles are just estimates. For example, the exact value of the sine of 60 degrees is $\frac{\sqrt{3}}{2}$. However, because radicals of numbers that aren't perfect squares are irrational and have an endless decimal value, a calculator carries that value out to a certain number of decimal places and then rounds it off. In this case, $\frac{\sqrt{3}}{2} \approx 0.8660254038$. This decimal has many more places than you usually need — normally, three or four decimal places is enough.

In trig, you frequently use the exact values of the most favorite angles because they give better results in computations and applications, so memorizing those exact values is a good idea.

The process for constructing a table of trig function values, which I explain in this section, is easy to remember, so you can create one quickly when you need to — either on paper or in your head.

A quick table for the three basic trig functions

The angles used most often in trig have trig functions with convenient exact values. Other angles don't cooperate anywhere near as nicely as these popular ones do.

A quick, easy way to memorize the exact trig-function values of the most common angles is to construct a table, starting with the sine function and working with a pattern of fractions and radicals. Create a table with the top row listing the angles, as shown in Figure 7-6. The first function, in the next row, is sine.

Figure 7-6: Constructing a table of exact values.

$\theta°$	0°	30°	45°	60°	90°
$\sin \theta$					

The entries following sin θ in the second row are the fractions and radicals with the following pattern:

✔ Each fraction has a denominator of 2.

✔ The numerators of the fractions are radicals with 0, 1, 2, 3, and 4 under them, in that order, as shown in Figure 7-7.

Figure 7-7:
Creating entries for the second row of the table of exact values.

θ°	0°	30°	45°	60°	90°
sin θ	$\frac{\sqrt{0}}{2}$	$\frac{\sqrt{1}}{2}$	$\frac{\sqrt{2}}{2}$	$\frac{\sqrt{3}}{2}$	$\frac{\sqrt{4}}{2}$

Next, simplify the fractions that can be simplified so the table becomes what you see in Figure 7-8:

$$\frac{\sqrt{0}}{2} = 0, \quad \frac{\sqrt{1}}{2} = \frac{1}{2}, \quad \frac{\sqrt{2}}{2} = \frac{\sqrt{2}}{2}, \quad \frac{\sqrt{3}}{2} = \frac{\sqrt{3}}{2}, \quad \frac{\sqrt{4}}{2} = 1$$

Figure 7-8:
The first row of the table with simplified values.

θ°	0°	30°	45°	60°	90°
sin θ	0	$\frac{1}{2}$	$\frac{\sqrt{2}}{2}$	$\frac{\sqrt{3}}{2}$	1

The next row, for the cosine, is just the sine's row in reverse order, as Figure 7-9 shows. This happens because you have the angles and their complements in reverse order, too.

Figure 7-9:
Adding the cosine row.

θ°	0°	30°	45°	60°	90°
sin θ	0	$\frac{1}{2}$	$\frac{\sqrt{2}}{2}$	$\frac{\sqrt{3}}{2}$	1
cos θ	1	$\frac{\sqrt{3}}{2}$	$\frac{\sqrt{2}}{2}$	$\frac{1}{2}$	0

The next row is for the tangent. In a right triangle, you find the tangent of an acute angle with the ratio $\frac{\text{opposite}}{\text{adjacent}}$ (refer to "The tangent function: Opposite over adjacent," earlier in this chapter). You get the same ratio when you divide sine by cosine. Here's how it works:

$$\frac{\text{sine}}{\text{cosine}} = \frac{\dfrac{\text{opposite}}{\text{hypotenuse}}}{\dfrac{\text{adjacent}}{\text{hypotenuse}}} = \frac{\text{opposite}}{\cancel{\text{hypotenuse}}} \cdot \frac{\cancel{\text{hypotenuse}}}{\text{adjacent}} = \frac{\text{opposite}}{\text{adjacent}}$$

Because you already know the values for sine and cosine, you can use this property (tangent equals sine divided by cosine) to get the tangent values for the table:

- ✔ For the tangent of 0 degrees, $\frac{0}{1} = 0$.

- ✔ The tangent of 30 degrees is $\dfrac{\frac{1}{2}}{\frac{\sqrt{3}}{2}} = \frac{1}{2} \cdot \frac{2}{\sqrt{3}} = \frac{1}{\sqrt{3}} = \frac{\sqrt{3}}{3}$.

- ✔ The tangent of 45 degrees is $\dfrac{\frac{\sqrt{2}}{2}}{\frac{\sqrt{2}}{2}} = 1$.

- ✔ The tangent of 60 degrees is $\dfrac{\frac{\sqrt{3}}{2}}{\frac{1}{2}} = \frac{\sqrt{3}}{2} \cdot \frac{2}{1} = \frac{\sqrt{3}}{1} = \sqrt{3}$.

- ✔ The tangent of 90 degrees is $\frac{1}{0}$, which is undefined. So, the tangent of 90 degrees doesn't have a value — it simply doesn't exist.

See Figure 7-10 for the completed table with the tangent row.

$\theta°$	0°	30°	45°	60°	90°
$\sin \theta$	0	$\frac{1}{2}$	$\frac{\sqrt{2}}{2}$	$\frac{\sqrt{3}}{2}$	1
$\cos \theta$	1	$\frac{\sqrt{3}}{2}$	$\frac{\sqrt{2}}{2}$	$\frac{1}{2}$	0
$\tan \theta$	0	$\frac{\sqrt{3}}{3}$	1	$\sqrt{3}$	undefined

Figure 7-10: The tangent row comes next.

A quick table for the three reciprocal trig functions

If you read "Taking It a Step Further: Reciprocal Functions," earlier in this chapter, you know that the reciprocal functions have values that are *reciprocals,* or flips, of the values for their respective functions. The reciprocal of sine is cosecant, so each function value for cosecant is the reciprocal of sine's. The same goes for the other two reciprocal functions. The table in Figure 7-11 shows the reciprocal in each case, in their simplified forms. Whenever you see *undefined,* it's because the original function value was 0, and the reciprocal of 0 has no value.

Figure 7-11:
The recip-
rocal
functions.

$\theta°$	0°	30°	45°	60°	90°
$\csc \theta$	undefined	2	$\sqrt{2}$	$\dfrac{2\sqrt{3}}{3}$	1
$\sec \theta$	1	$\dfrac{2\sqrt{3}}{3}$	$\sqrt{2}$	2	undefined
$\cot \theta$	undefined	$\sqrt{3}$	1	$\dfrac{\sqrt{3}}{3}$	0

Refer to Chapter 8 for the trig values of angles measuring more than 90 degrees. These values are developed using the chart created here.

Chapter 8

Trading Triangles for Circles: Circular Functions

*O*ne of the ways that mathematicians first defined the trig functions was by using ratios formed from the measures of the sides of right triangles (see Chapter 7). Right triangles and the measures of their sides are convenient and easy to construct. This fact led to a sort of natural development of the trig functions, and it proved to be most useful because it allowed engineers, astronomers, and mathematicians to make accurate calculations of the heights of tall objects, areas of large expanses, and predictions of eclipses and other astronomical phenomena. But, of course, they couldn't stop there. The world of trigonometry and its applications opened up even more when they expanded the trig functions and properties to angles of *any* measure — positive and negative — not just those limited to a right triangle. This extension of the angles allowed them to calculate the areas of triangles containing obtuse angles and conduct navigational plots. The best place to begin describing these new function values and comparing them with the old is with the most basic of all circles — the unit circle.

Getting Acquainted with the Unit Circle

The *unit circle* is a circle with its center at the origin of the coordinate plane and with a radius of 1 unit. Any circle with its center at the origin has the equation $x^2 + y^2 = r^2$, where r is the radius of the circle. In the case of a unit circle, the equation is $x^2 + y^2 = 1$. This equation shows that the points lying on the unit circle have to have coordinates (*x*- and *y*-values) that, when you square each of them and then add those values together, equal 1. The coordinates for the points lying on the unit circle and also on the axes are (1,0), (–1,0), (0,1), and (0,–1). These four points (called *intercepts*) are shown in Figure 8-1.

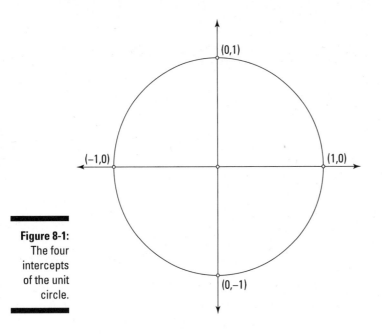

Figure 8-1:
The four
intercepts
of the unit
circle.

Placing points on the unit circle

The rest of the points on the unit circle aren't as nice and neat as those you see in Figure 8-1. They all have fractions or radicals — or both — in them. For instance, the point $\left(\frac{1}{2}, \frac{\sqrt{3}}{2}\right)$ lies on the unit circle. Look at how these coordinates work in the equation of the unit circle:

$$\left(\frac{1}{2}\right)^2 + \left(\frac{\sqrt{3}}{2}\right)^2 = \frac{1}{4} + \frac{3}{4} = 1$$

When you square each coordinate and add those values together, you get 1.

Any combination of these two coordinates, whether the coordinates are positive or negative, gives you a different point on the unit circle. They all work because whether a number is positive or negative, its square is the same positive number. Here are some combinations of those two coordinates that satisfy the unit-circle equation:

$$\left(\frac{1}{2}, \frac{\sqrt{3}}{2}\right) \qquad \left(-\frac{1}{2}, \frac{\sqrt{3}}{2}\right) \qquad \left(\frac{1}{2}, -\frac{\sqrt{3}}{2}\right)$$

$$\left(-\frac{\sqrt{3}}{2}, \frac{1}{2}\right) \qquad \left(-\frac{\sqrt{3}}{2}, -\frac{1}{2}\right)$$

Another pair of coordinates that works on the unit circle is $\left(\frac{\sqrt{2}}{2}, \frac{\sqrt{2}}{2}\right)$, because the sum of the squares is equal to 1:

$$\left(\frac{\sqrt{2}}{2}\right)^2 + \left(\frac{\sqrt{2}}{2}\right)^2 = \frac{2}{4} + \frac{2}{4} = \frac{4}{4} = 1$$

The numbers that continually crop up as coordinates of points on the unit circle are $0, \frac{1}{2}, \frac{\sqrt{2}}{2}, \frac{\sqrt{3}}{2}, 1$. If you read Chapter 7, they should look familiar — they're the sine and cosine values of the most common acute-angle measures. Figure 8-2 shows the locations of those points on the unit circle.

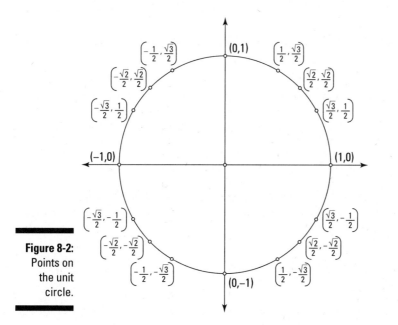

Figure 8-2:
Points on the unit circle.

Babylonian mathematics

The Babylonians were of an ancient culture that had a good deal of influence on the development of mathematics in many areas:

- **They developed a system of written symbols.** Way back in the day, the Babylonians developed a form of writing that was based on wedge-shaped symbols called *cuneiform*. Their work has been preserved in the clay tablets that they wrote on, but this way of writing was really cumbersome, so they couldn't write very fast or for very long.

- **They gave us the time of day.** The Babylonians divided the day into 24 hours, each hour into 60 minutes, and each minute into 60 seconds. This division resulted in their base-60 counting system — called the *hexasegimal* system — which means that their number system was base 60 and had different characters for 1; 10; 60; 600; 3,600;

36,000; and 216,000. Their system allowed them to easily write fractions.

- **They beat Pythagoras to the punch.** These ancients had a knowledge of trigonometry and the Pythagorean theorem 1,000 years before Pythagoras did; they just didn't get the credit he did.

- **They were neat freaks.** Their methods for solving problems were very logical and systematic. They preferred orderly procedures based on tables and facts, which is probably why they knew about π and could approximate its value.

The Babylonians thought in terms of algebra and trigonometry, but you probably wouldn't recognize their notes if you were to pick up a tablet or two to read what they discovered. You'll find this book much more readable!

The points on the unit circle shown in Figure 8-2 are frequently used in trigonometry and other math applications, but they aren't the only points on that circle. Every circle has an infinite number of points with all sorts of interesting coordinates — even more interesting than those already shown. If you're looking for the coordinates of some other point on the unit circle, you can just pick some number between –1 and 1 to be the x- or the y-value and then solve for the other value. I describe this method for finding the other part of a coordinate in the next section. All these other coordinates come into play when you're drawing a ray that starts at the unit circle's center and want to find the trig functions of the angle formed by that ray and the positive x-axis.

Finding a missing coordinate

If you have the value of one of a point's coordinates on the unit circle and need to find the other, you can substitute the known value into the unit-circle equation and solve for the missing value.

You can choose any number between 1 and –1, because that's how far the unit circle extends along the *x*- and *y*-axes. For example, say $\frac{2}{5}$ is the *x*-coordinate of a point on the unit circle. You can find the *y*-coordinate like so:

1. **Substitute the *x*-coordinate value into the unit-circle equation.**

$$\left(\frac{2}{5}\right)^2 + y^2 = 1$$

2. **Square the *x*-coordinate and subtract that value from each side.**

$$\frac{4}{25} + y^2 = 1$$
$$y^2 = 1 - \frac{4}{25} = \frac{21}{25}$$

3. **Take the square root of each side.**

$$\sqrt{y^2} = \pm\sqrt{\frac{21}{25}}$$
$$y = \pm\frac{\sqrt{21}}{5}$$

Note that the *y*-coordinate can have two values, because the unit circle has two different points for every particular *x*-coordinate (and for every *y*-coordinate). Look at Figure 8-3, and you can see how that happens.

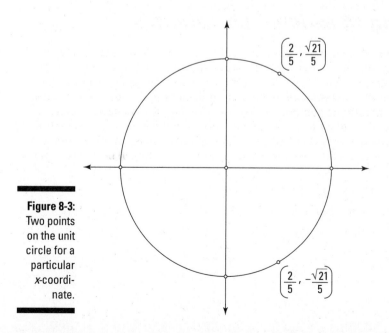

Figure 8-3: Two points on the unit circle for a particular *x*-coordinate.

$$\left(\frac{2}{5}, \frac{\sqrt{21}}{5}\right)$$

$$\left(\frac{2}{5}, -\frac{\sqrt{21}}{5}\right)$$

Another example: Find the x-coordinate (or coordinates) if the y-coordinate is $-\frac{7}{25}$.

1. **Substitute the y-coordinate value into the unit-circle equation.**

$$x^2 + \left(-\frac{7}{25}\right)^2 = 1$$

2. **Square the y-coordinate and subtract that value from each side.**

$$x^2 + \frac{49}{625} = 1$$
$$x^2 = 1 - \frac{49}{625}$$
$$= \frac{576}{625}$$

3. **Take the square root of each side.**

$$\sqrt{x^2} = \pm\sqrt{\frac{576}{625}}$$
$$x = \pm\frac{24}{25}$$

As you can see, the x-coordinate here has two values, and the two points are $\left(-\frac{7}{25}, \frac{24}{25}\right)$ and $\left(-\frac{7}{25}, -\frac{24}{25}\right)$.

Sticking to rational coordinates

You may have noticed, in the last section, that one problem resulted in a coordinate with a radical in it and the other didn't. Radicals can't be avoided (as you read in the news) when doing trig problems, but sometimes you just need to keep things rational. A *rational* number is a real number that can be written as a fraction. And rational numbers have decimal values that behave — unlike the decimal values of radicals (irrational numbers). What I have for you here is a way of assuring yourself that you'll get only rational coordinates for a point on the unit circle. To do this, use the following formula, letting m be any rational number:

$$(x,y) = \left(\frac{1-m^2}{1+m^2}, \frac{2m}{1+m^2}\right)$$

Want to see the formula at work? First, I show you the formula starting with a nice, civilized rational number; let $m = 4$.

1. Replace each m in the formula with 4.

$$\left(\frac{1-m^2}{1+m^2}, \frac{2m}{1+m^2}\right) = \left(\frac{1-4^2}{1+4^2}, \frac{2\cdot 4}{1+4^2}\right)$$

2. Simplify.

$$= \left(\frac{1-16}{1+16}, \frac{8}{1+16}\right) = \left(-\frac{15}{17}, \frac{8}{17}\right)$$

Skeptical? Just check the coordinates in the equation for the unit circle.

$$\left(-\frac{15}{17}\right)^2 + \left(\frac{8}{17}\right)^2 = \frac{225}{289} + \frac{64}{289} = \frac{289}{289} = 1$$

And now, to really convince even the biggest skeptics, I choose $m = -\frac{2}{3}$.

1. Replace each m in the formula with $-\frac{2}{3}$.

$$\left(\frac{1-m^2}{1+m^2}, \frac{2m}{1+m^2}\right) = \left(\frac{1-\left(-\frac{2}{3}\right)^2}{1+\left(-\frac{2}{3}\right)^2}, \frac{2\left(-\frac{2}{3}\right)}{1+\left(-\frac{2}{3}\right)^2}\right)$$

2. Simplify.

$$= \left(\frac{1-\frac{4}{9}}{1+\frac{4}{9}}, \frac{-\frac{4}{3}}{1+\frac{4}{9}}\right) = \left(\frac{\frac{5}{9}}{\frac{13}{9}}, \frac{-\frac{4}{3}}{\frac{13}{9}}\right) = \left(\frac{5}{13}, -\frac{12}{13}\right)$$

And, of course, this point checks, too.

I'm going to let you in on a little secret: This formula is based on using Pythagorean triples in the numerators and denominators of the fractions. There's a bit more to it — creating the formula using a slope of a line through an intercept of the unit circle — but you don't need all that to take advantage of the convenience of the numbers produced by the formula.

Going Full Circle with the Angles

The unit circle is a platform for describing all the possible angle measures from 0 to 360 degrees, all the negatives of those angles, plus all the multiples of the positive and negative angles from negative infinity to positive infinity. In other words, the unit circle shows you all the angles that exist. Because a right triangle can only measure angles of 90 degrees or less, the circle allows for a much-broader range.

Staying positive

The positive angles on the unit circle are measured with the initial side on the positive x-axis and the terminal side moving counterclockwise around the origin (to figure out which side is which, see Chapter 4). Figure 8-4 shows some positive angles labeled in both degrees and radians.

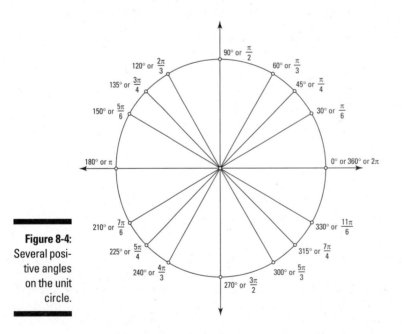

Figure 8-4:
Several positive angles on the unit circle.

In Figure 8-4, notice that the terminal sides of the angles measuring 30 degrees and 210 degrees, 60 degrees and 240 degrees, and so on form straight lines. This fact is to be expected because the angles are 180 degrees apart, and a straight angle measures 180 degrees. You see the significance of this fact when you deal with the trig functions for these angles in Chapter 9.

Being negative or multiplying your angles

Just when you thought that angles measuring up to 360 degrees or 2π radians was enough for anyone, you're confronted with the reality that many of the basic angles have negative values and even multiples of themselves. If you measure angles clockwise instead of counterclockwise, then the angles have negative measures: A 30-degree angle is the same as an angle measuring –330 degrees, because they have the same terminal side. Likewise, an angle of $\frac{5\pi}{3}$ is the same as an angle of $-\frac{\pi}{3}$. For rules on how to change degree measure to radian measure, refer to Chapter 1.

But wait — you have even more ways to name an angle. By doing a complete rotation of two (or more) and adding or subtracting 360 degrees or a multiple of it before settling on the angle's terminal side, you can get an infinite number of angle measures, both positive and negative, for the same basic angle. For example, an angle of 60 degrees has the same terminal side as that of a 420-degree angle and a –300-degree angle. Figure 8-5 shows many names for the same 60-degree angle in both degrees and radians.

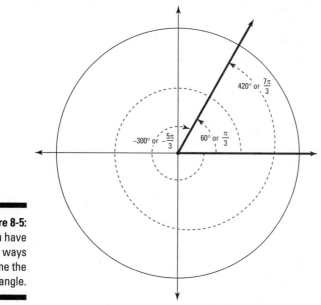

Figure 8-5:
You have
many ways
to name the
same angle.

Although this name-calling of angles may seem pointless at first, there's more to it than arbitrarily using negatives or multiples of angles just to be difficult. The angles that are related to one another have trig functions that are also related, if not the same (more on that in Chapter 9).

Locating and computing reference angles

Each of the angles in a unit circle has a *reference angle,* which is always a positive acute angle (except the angles that are already positive and acute). By identifying the reference angle, you can determine the function values for that reference angle and, ultimately, the original angle. Usually, solving for the reference angle first is much easier than trying to determine a trig function for the original angle. The trig functions have values that repeat over and over; sometimes those values are positive, and sometimes they're negative. Using a reference angle helps keep the number of different values to a minimum. You just assign the positive or negative sign after determining a numerical value for the function from the reference angle.

You determine a reference angle by looking at the terminal side of the angle you're working with and its relation with the positive or negative *x*-axis (depending on which quadrant the terminal side is in). The following tells you how to measure the reference angle when you're given the terminal side of the angle:

 ✔ **Quadrant I (QI):** The reference angle is the same as the original angle itself.

 ✔ **Quadrant II (QII):** The reference angle is the measure from the terminal side down to the negative *x*-axis.

 ✔ **Quadrant III (QIII):** The reference angle is the measure from the negative *x*-axis down to the terminal side.

 ✔ **Quadrant IV (QIV):** The reference angle is the measure from the terminal side up to the positive *x*-axis.

Figure 8-6 shows the positions of the reference angles in the four quadrants.

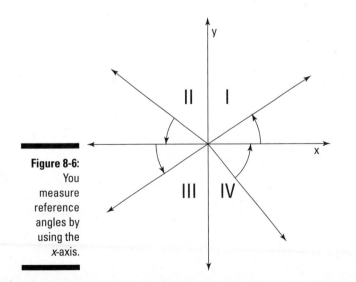

Figure 8-6: You measure reference angles by using the *x*-axis.

As with all angles, you measure reference angles in degrees or radians. I have to admit that I sometimes prefer to work in degrees and will convert a radian measure to do these computations. Whichever method you choose is fine — it's all a matter of taste.

Figuring the angle measure in degrees

To compute the measure (in degrees) of the reference angle for any given angle θ, use the rules in Table 8-1.

Table 8-1	Finding Reference Angles in Degrees	
Quadrant	*Measure of Angle θ*	*Measure of Reference Angle*
I	0° to 90°	θ
II	90° to 180°	$180° - \theta$
III	180° to 270°	$\theta - 180°$
IV	270° to 360°	$360° - \theta$

Using Table 8-1, find the reference angle for 200 degrees:

1. **Determine the quadrant in which the terminal side lies.**

   A 200-degree angle is between 180 and 270 degrees, so the terminal side is in QIII.

2. **Do the operation indicated for that quadrant.**

   Subtract 180 degrees from the angle, which is 200 degrees. You find that $200 - 180 = 20$, so the reference angle is 20 degrees.

Sometimes angle measures don't fit neatly in the ranges shown in Table 8-1. For example, you may need to find the reference angle for a negative angle or a multiple of an angle.

To find the reference angle for –340 degrees:

1. **Determine the quadrant in which the terminal side lies.**

   A –340-degree angle is equivalent to a 20-degree angle. (You get the positive angle measure by adding 360, or one full revolution around the origin, to the negative measure.) A 20-degree angle has its terminal side in QI.

2. **Do the operation indicated for that quadrant.**

   Angles in the first quadrant are their own reference angle, so the reference angle is 20 degrees.

Figuring the angle measure in radians

To compute the measure (in radians) of the reference angle for any given angle θ, use the rules in Table 8-2.

Table 8-2	Finding Reference Angles in Radians	
Quadrant	**Measure of Angle θ**	**Measure of Reference Angle**
I	0 to $\frac{\pi}{2}$	θ
II	$\frac{\pi}{2}$ to π	$\pi - \theta$
III	π to $\frac{3\pi}{2}$	$\theta - \pi$
IV	$\frac{3\pi}{2}$ to 2π	$2\pi - \theta$

To find the reference angle for $\frac{15\pi}{16}$, for example:

1. **Determine the quadrant in which the terminal side lies.**

   An angle measuring $\frac{15\pi}{16}$ has its terminal side in QII, which you know because $\frac{15}{16}$ is slightly less than 1, making the angle slightly less than π.

2. **Do the operation indicated for that quadrant.**

   Subtract $\frac{15\pi}{16}$ from π. When you do so, you get $\pi - \frac{15\pi}{16} = \frac{16\pi}{16} - \frac{15\pi}{16} = \frac{\pi}{16}$, so the reference angle is $\frac{\pi}{16}$.

Chapter 9

Defining Trig Functions Globally

. . . .

. . . .

*T*he six basic trig functions all had humble beginnings with the right triangle and its angles. The unit circle opens up a whole new world
for the input values into those functions. Because of the nature of trig
functions — they repeat the same patterns over and over — the output
values show up regularly. This repetition is a good thing; you recognize
where in the pattern a particular input belongs and then assign the output.
Life is good.

Defining Trig Functions for All Angles

So many angles are used in trigonometry and other math areas, and the
majority of those angles are multiples of 30 and 45 degrees. So, having a trick
up your sleeve letting you quickly access the function values of this frequent-
flier list of angles makes perfect sense. All you need to know or memorize
are the values of the trig functions for 0-, 30-, 45-, 60-, and 90-degree angles in
order to determine all the trig functions of *all* the angles, positive or negative,
that are multiples of 30 or 45 degrees, which are the two most basic, founda-
tional angles. Finding these function values for a particular angle is a three-
step process: (1) Find the measure of the angle's reference angle, (2) Assign
the correct numerical value, and (3) Determine whether the function value is
positive or negative.

Putting reference angles to use

The first step to finding the function value of one of the angles that's a multiple of 30 or 45 degrees is to find the reference angle. When the reference angle comes out to be 0, 30, 45, 60, or 90 degrees, you can use the function value of that angle and then figure out the sign (see the next section). Use Table 8-1 or Table 8-2 to find the reference angle.

All angles with a 30-degree reference angle have trig functions whose absolute values are the same as those of the 30-degree angle. The sines of 30, 150, 210, and 330 degrees, for example, are all either $\frac{1}{2}$ or $-\frac{1}{2}$. Likewise, using a 45-degree angle as a reference angle, the cosines of 45, 135, 225, and 315 degrees, for example, are all $\frac{\sqrt{2}}{2}$ or $-\frac{\sqrt{2}}{2}$.

Labeling the optimists and pessimists

The sine values for 30, 150, 210, and 330 degrees are $\frac{1}{2}, \frac{1}{2}, -\frac{1}{2}$, and $-\frac{1}{2}$, respectively. All these multiples of 30 degrees have an absolute value of $\frac{1}{2}$ (as I explain in the preceding section). The following rule and Figure 9-1 help you determine whether a trig-function value is positive or negative. First, note that each quadrant in the figure is labeled with a letter. The letters aren't random; they stand for trig functions.

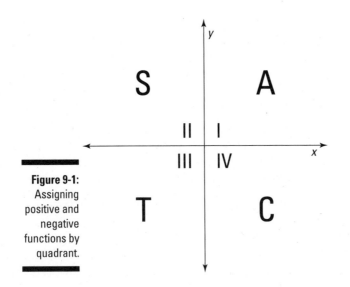

Figure 9-1:
Assigning
positive and
negative
functions by
quadrant.

Reading around the quadrants, starting with QI and going counterclockwise, the rule goes like this: If the terminal side of the angle is in the quadrant with letter

> ✔ **A:** All functions are positive.
>
> ✔ **S:** Sine and its reciprocal, cosecant, are positive.
>
> ✔ **T:** Tangent and its reciprocal, cotangent, are positive.
>
> ✔ **C:** Cosine and its reciprocal, secant, are positive.

In QII, only sine and cosecant are positive. All the other function values for angles in this quadrant are negative — and the rule continues in like fashion for the other quadrants.

My trig teacher, Dr. Johnson, showed me a great way to remember this rule: "All Students Take Calculus." (And so I did!) If math is already giving you nightmares, maybe you'd prefer "Any Snake Teases Chickens" or "Apple Sauce Turns Colors." Make up your own! Have at it!

Combining all the rules

Using the rules for reference angles, the values of the functions of certain acute angles (see Chapter 7), and the rule for the signs of the functions, you can determine the trig functions for any angles found on the unit circle — any that are graphed in *standard position* (meaning the vertex of the angle is at the origin, and the initial side lies along the positive *x*-axis). Figure 9-2 combines information from this chapter and Chapter 8 to give you the information you need.

Now, armed with all the necessary information, find the tangent of 300 degrees.

1. **Find the reference angle.**

   Using the top chart in Figure 9-2, you can see that a 300-degree angle has its terminal side in the fourth quadrant, so you find the reference angle by subtracting 300 from 360. Therefore, the measure of the reference angle is 60 degrees.

2. **Find the numerical value of the tangent.**

   Using the middle chart in Figure 9-2, you see that the numerical value of the tangent of 60 degrees is $\sqrt{3}$.

3. **Find the sign of the tangent.**

   Because a 300-degree angle is in the fourth quadrant, and angles in that quadrant have negative tangents (refer to the preceding section), the tangent of 300 degrees is $-\sqrt{3}$.

Quadrant	Measure of Angle θ	Measure of Reference Angle
I	0° to 90°	θ
II	90° to 180°	$180° - \theta$
III	180° to 270°	$\theta - 180°$
IV	270° to 360°	$360° - \theta$

$\theta°$	0°	30°	45°	60°	90°
$\sin \theta$	0	$\dfrac{1}{2}$	$\dfrac{\sqrt{2}}{2}$	$\dfrac{\sqrt{3}}{2}$	1
$\cos \theta$	1	$\dfrac{\sqrt{3}}{2}$	$\dfrac{\sqrt{2}}{2}$	$\dfrac{1}{2}$	0
$\tan \theta$	0	$\dfrac{\sqrt{3}}{3}$	1	$\sqrt{3}$	undefined

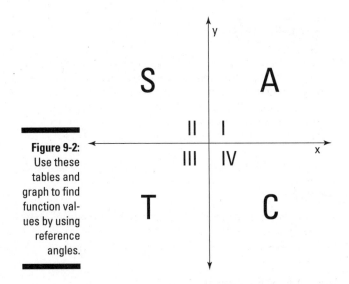

Figure 9-2:
Use these tables and graph to find function values by using reference angles.

To try your hand at working with radians, find the cosecant of $\dfrac{7\pi}{6}$.

1. **Find the reference angle.**

    To use the top chart in Figure 9-2, you need to determine the degree equivalence for an angle measuring $\dfrac{7\pi}{6}$. In Chapter 5, you find the formula for converting from radians to degrees. Using the formula in this case, you get that $\frac{7\pi}{6}$ is equivalent to 210°. This angle is in the third quadrant, so, going back to radians, you find the reference angle by subtracting π from $\frac{7\pi}{6}$, resulting in $\frac{7\pi}{6} - \pi = \frac{\pi}{6}$.

2. **Find the numerical value of the cosecant.**

   In the middle chart of Figure 9-2, the cosecant doesn't appear. However, the reciprocal of the cosecant is sine. So find the value of the sine, and use its reciprocal. The sine of $\frac{\pi}{6}$ is $\frac{1}{2}$, which means that the cosecant of $\frac{\pi}{6}$ is 2 (the reciprocal).

3. **Find the sign of the cosecant.**

   In the third quadrant, the cosecant of an angle is negative (refer to the preceding section), so the cosecant of $\frac{7\pi}{6}$ is –2.

Using Coordinates of Circles to Solve for Trig Functions

Another way to find the values of the trig functions for angles is to use the coordinates of points on a circle that has its center at the origin. Letting the positive x-axis be the initial side of an angle, you can use the coordinates of the point where the terminal side intersects with the circle to determine the trig functions. Figure 9-3 shows a circle with a radius of r that has an angle drawn in standard position.

The equation of a circle is $x^2 + y^2 = r^2$ (flip back to Chapter 2 for a refresher). Based on this equation and the coordinates of the point (x,y), where the terminal side of the angle intersects the circle, the six trig functions for angle θ are defined as follows:

$$\sin\theta = \frac{y}{r} \qquad\qquad \csc\theta = \frac{r}{y}$$

$$\cos\theta = \frac{x}{r} \qquad\qquad \sec\theta = \frac{r}{x}$$

$$\tan\theta = \frac{y}{x} \qquad\qquad \cot\theta = \frac{x}{y}$$

You can see where these definitions come from if you picture a right triangle formed by dropping a perpendicular segment from the point (x,y) to the x-axis. Figure 9-4 shows such a right triangle. Remember that the x-value is to the right (or left) of the origin, and the y-value is above (or below) the x-axis — and use those values as lengths of the triangle's sides. Therefore, the side oppo-site angle θ is y, the value of the y-coordinate. The adjacent side is x, the value of the x-coordinate.

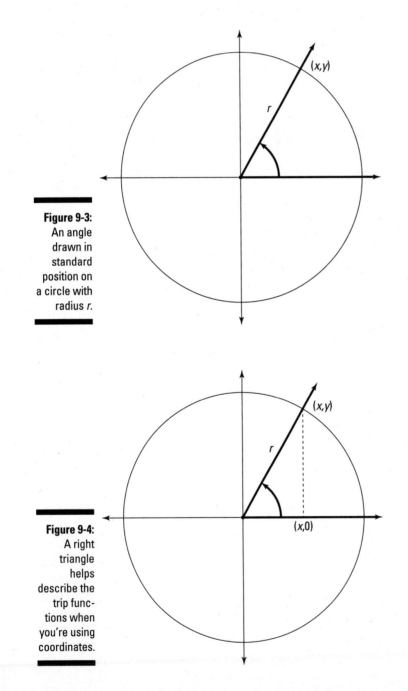

Figure 9-3:
An angle
drawn in
standard
position on
a circle with
radius r.

Figure 9-4:
A right
triangle
helps
describe the
trip func-
tions when
you're using
coordinates.

Take note that for angles in the second quadrant, for example, the *x*-values are negative, and the *y*-values are positive. The radius, however, is always a positive number. With the *x*-values negative and the *y*-values positive, using the definitions for the functions listed earlier in this section, you see that the sine and cosecant are positive, but the other functions are all negative, because they all have an *x* in their ratios. The signs of the trig functions all fall into line when you use this coordinate system, so no need to worry about remembering the ASTC rule here. (For more on that rule and when to use it, see "Labeling the optimists and pessimists," earlier in this chapter.)

Calculating with coordinates on the unit circle

Calculating trig functions of angles within a unit circle is easy as pie. Figure 9-5 shows a unit circle, which has the equation $x^2 + y^2 = 1$, along with some points on the circle and their coordinates.

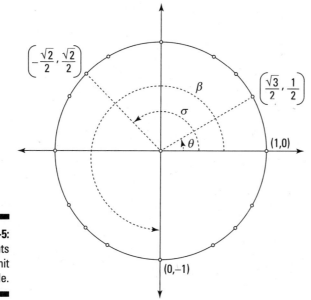

Figure 9-5: Some points on the unit circle.

Using the angles in Figure 9-5, find the tangent of θ.

1. **Find the x- and y-coordinates of the point where the angle's terminal side intersects with the circle.**

   The coordinates are $x = \frac{\sqrt{3}}{2}$ and $y = \frac{1}{2}$. The radius is $r = 1$.

2. **Determine the ratio for the function and substitute in the values.**

   The ratio for the tangent is $\frac{y}{x}$, so you find that

   $$\frac{y}{x} = \frac{\frac{1}{2}}{\frac{\sqrt{3}}{2}} = \frac{1}{2} \cdot \frac{2}{\sqrt{3}} = \frac{1}{\sqrt{3}} \cdot \frac{\sqrt{3}}{\sqrt{3}} = \frac{\sqrt{3}}{3}$$

Next, using the angles in Figure 9-5, find the cosine of σ.

1. **Find the x- and y-coordinates of the point where the terminal side of the angle intersects with the circle.**

   The coordinates are $x = -\frac{\sqrt{2}}{2}$ and $y = \frac{\sqrt{2}}{2}$; the radius is $r = 1$.

2. **Determine the ratio for the function and substitute in the values.**

   The ratio for the cosine is $\frac{x}{r}$, which means that you need only the

   x-coordinate, so $\frac{x}{r} = \frac{-\frac{\sqrt{2}}{2}}{1} = -\frac{\sqrt{2}}{2}$.

Now, using the angles in Figure 9-5, find the cosecant of β.

1. **Find the x- and y-coordinates of the point where the terminal side of the angle intersects with the circle.**

   The coordinates are $x = 0$ and $y = -1$; the radius is $r = 1$.

2. **Determine the ratio for the function and substitute in the values.**

   The ratio for cosecant is $\frac{r}{y}$, which means that you need only the

   y-coordinate, so $\frac{r}{y} = \frac{1}{-1} = -1$.

Calculating with coordinates on any circle at the origin

You don't need a unit circle to use this coordinate business when determining the function values of angles graphed in standard position on a circle. You can use a circle with any radius, as long as the center is at the origin. The standard equation for a circle centered at the origin is $x^2 + y^2 = r^2$.

Using the angles in Figure 9-6, find the sine of α.

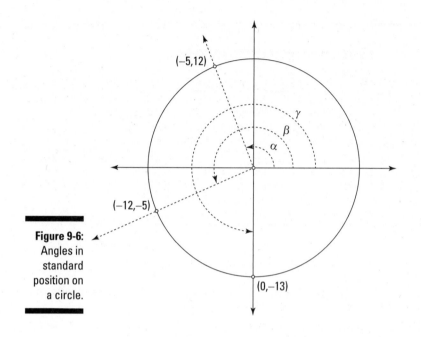

Figure 9-6:
Angles in
standard
position on
a circle.

1. **Find the *x*- and *y*-coordinates of the point where the terminal side of the angle intersects with the circle.**

   The coordinates are $x = -5$ and $y = 12$.

2. **Determine the radius of the circle.**

   The equation of the circle is $x^2 + y^2 = r^2$. Replacing the x and y in this equation with -5 and 12, respectively, you get $(-5)^2 + (12)^2 = 25 + 144 = 169 = r^2$. The square root of 169 is 13, so the radius is 13.

3. **Determine the ratio for the function, and substitute in the values.**

   The ratio for sine is $\frac{y}{r}$, which means that you need only the y-coordinate and radius, so $\frac{y}{r} = \frac{12}{13}$.

Next, using the angles in Figure 9-6, find the cotangent of β.

1. **Find the *x*- and *y*-coordinates of the point where the terminal side of the angle intersects with the circle.**

   The coordinates are $x = -12$ and $y = -5$.

   The cotangent function uses only the x- and y-coordinates, so you don't need to solve for the radius.

2. **Determine the ratio for the function, and substitute in the values.**

   The ratio of cotangent is $\frac{x}{y}$, so $\frac{x}{y} = \frac{-12}{-5} = \frac{12}{5}$.

Ancient math contest

Even in the late 1500s, mathematicians around the world found themselves competing with one another. The Belgian mathematician Adriaan Van Roomen challenged other mathematicians to solve a polynomial equation of the 45th degree. It looked something like this (with variables and numbers in place of the ellipsis, of course): $x^{45} - 45x^{43} + 945x^{41} - \ldots + 45x = k$, where k is some constant. At that time,

the mathematicians in Belgium and France found themselves in quite a competition, because Van Roomen suggested that no one in France would be able to solve the equation. But French mathematician François Viète put him to shame: He used trigonometry to solve the puzzle. (He let k be equal to the sine of 45 degrees and applied trig identities to find the positive solutions.)

Now, using the angles in Figure 9-6, find the secant of γ.

1. **Find the x- and y-coordinates of the point where the terminal side of the angle intersects with the circle.**

   The coordinates are $x = 0$ and $y = -13$.

2. **Determine the radius of the circle.**

   Per the first example in this section, the radius is 13.

3. **Determine the ratio for the function, and substitute in the values.**

   The ratio for secant is $\frac{r}{x}$, so you need only the x-coordinate; substituting in, you get $\frac{r}{x} = \frac{13}{0}$. This answer is undefined, which means that angle γ has no secant. For the reason that γ has no secant, refer to the next section.

Defining Domains and Ranges of Trig Functions

The *domain* of a function consists of all the input values that a function can handle — the way the function is defined. Of course, you want to get output values (which make up the *range*) when you enter input values (for the basics on domain and range, see Chapter 3). But sometimes, when you input something that doesn't belong in the function, you end up with some impossible situations. In these cases, you need to limit what you put into the function — the domain has to be restricted. For example, the cosecant is defined as the hypotenuse divided by the opposite side (see Chapter 7). If the terminal side of the angle is on the *x*-axis, then the opposite side is 0, and you're asked to divide by 0. Impossible!

Trig functions have domains that are angle measures (the inputs are all angles), either in degrees or radians. The outputs of the trig functions are real numbers. The hitch here is that the different trig functions have different domains and ranges. You can't put just any angle into some of the functions. Sine and cosine are very cooperative and have the same domain and range. The tangent function and the reciprocal functions, however, all differ. The best way to describe these different domains and ranges is visually: Refer to the coordinate plane with a circle centered at the origin and a right triangle inside it, formed by dropping a line from any point (*x,y*) on the circle to the *x*-axis (see Figure 9-7). Remember that *r* stands for the radius of the circle (and also the hypotenuse of the right triangle in this figure). When that hypotenuse lies along one of the axes, one of the sides of the triangle is equal to 0, which is a no-no in the denominator of a fraction.

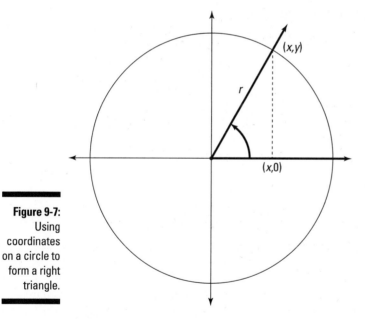

Figure 9-7:
Using
coordinates
on a circle to
form a right
triangle.

Consider the values of the variables in Figure 9-7 in relation to one another. The radius, *r*, is always positive. And the absolute values of *x* and *y* (the lengths of the segments they represent) are always smaller than *r*, unless the point (*x,y*) is on one of the axes — then one of the values is equal to *r* and the other is equal to 0.

Friendly functions: Sine and cosine

The sine and cosine functions are unique in the world of trig functions, because their ratios always have a value. No matter what angle you input, you get a resulting output. The value you get may be 0, but that's a number, too. In reference to the coordinate plane, sine is $\frac{y}{r}$, and cosine is $\frac{x}{r}$.

The radius, r, is always some positive number (which is why these functions always have a value, because they don't ask you to divide by 0), and r is always a number greater than (or equal to) the absolute value of x or y.

Domains of sine and cosine

The domains of sine and cosine are infinite. In trig speak, you say something like this: If θ represents all the angles in the domain of the two functions $f(\theta) = \sin \theta$ and $g(\theta) = \cos \theta$, then $-\infty < \theta < \infty$, which means that θ can be any angle in degrees or radians — any real number.

Ranges of sine and cosine

The output values for sine and cosine are always between (and including) -1 and 1. In trig speak, it goes something like this: If $f(\theta)$ and $g(\theta)$ represent the output values of the functions $f(\theta) = \sin \theta$ and $g(\theta) = \cos \theta$, then $-1 \le f(\theta) \le 1$ and $-1 \le g(\theta) \le 1$.

The ratios $\frac{y}{r}$ and $\frac{x}{r}$ will never be improper fractions — the numerator can never be greater than the denominator — because the value of r, the radius, is always the biggest number. At best, if the angle θ has a terminal side on an axis (meaning that one of the sides is equal to r), then the value of those ratios is 1 or -1.

Close cousins of their reciprocals: Cosecant and secant

The cosecant and secant functions are closely tied to sine and cosine, because they're the respective reciprocals. In reference to the coordinate plane, cosecant is $\frac{r}{y}$, and secant is $\frac{r}{x}$. The value of r is the length of the hypotenuse of a right triangle — which, as you find at the beginning of this section, is always positive and always greater than x and y. The only problem that arises when computing these functions is when either x or y is 0 — when the terminal side of the angle is on an axis. A function with a 0 in the denominator creates a number or value that doesn't exist (in math speak, the result is *undefined*), so anytime *x* or *y* is 0, you don't get any output from the cosecant or secant functions. The *x* is 0 when the terminal side is on the *y*-axis, and the *y* is 0 when the terminal side is on the *x*-axis.

Domains of cosecant and secant

The domains of cosecant and secant are restricted — you can only use the functions for angle measures with output numbers that exist.

Any time the terminal side of an angle lies along the *x*-axis (where *y* = 0), you can't perform the cosecant function on that angle. In trig speak, the rule looks like this: If $h(\theta) = \csc \theta$, then $\theta \neq 0, 180, 360, 540, \ldots$, or any multiple of 180 degrees. In radians, $\theta \neq 0, \pi, 2\pi, 3\pi, \ldots$, or any multiple of π.

Anytime the terminal side of an angle lies along the *y*-axis (where *x* = 0), you can't perform the secant function on that angle. So, in trig speak, you'd say this: If $k(\theta) = \sec \theta$, then $\theta \neq 90, 270, 450, 630, \ldots$, or any odd multiple of 90 degrees. In radians, $\theta \neq \frac{\pi}{2}, \frac{3\pi}{2}, \frac{5\pi}{2}, \frac{7\pi}{2}, \cdots$, or any odd multiple of $\frac{\pi}{2}$.

Ranges of cosecant and secant

The ratios of the cosecant and secant functions on the coordinate plane, $\frac{r}{y}$ and $\frac{r}{x}$, have the hypotenuse, *r*, in the numerator. Because *r* is always positive and greater than or equal to *x* and *y*, these fractions are always *improper* (greater than 1) or equal to 1. The ranges of these two functions never include *proper* fractions (numbers between –1 and 1).

If $h(\theta)$ and $k(\theta)$ are the output values of the functions $h(\theta) = \csc \theta$ and $k(\theta) = \sec \theta$, then $h(\theta) \leq -1$ or $h(\theta) \geq 1$ and $k(\theta) \leq -1$ or $k(\theta) \geq 1$.

Brothers out on their own: Tangent and cotangent

The tangent and cotangent are related not only by the fact that they're reciprocals, but also by the behavior of their ranges. In reference to the coordinate plane, tangent is $\frac{y}{x}$, and cotangent is $\frac{x}{y}$. The domains of both functions are restricted, because sometimes their ratios could have zeros in the denominator, but their ranges are infinite.

Domains of tangent and cotangent

Because *x* can't equal 0 for the tangent function to work, this rule holds true: If $m(\theta) = \tan \theta$, then $\theta \neq 90, 270, 450, 630, \ldots$, or any odd multiple of 90 degrees. In radians, $\theta \neq \frac{\pi}{2}, \frac{3\pi}{2}, \frac{5\pi}{2}, \frac{7\pi}{2}, \cdots$, or any odd multiple of $\frac{\pi}{2}$. Both the tangent and secant functions have ratios with x in the denominator, making their domains the same.

In order for the cotangent function to work, y can't equal 0. If $n(\theta) = \cot \theta$, then $\theta \neq 0, 180, 360, 540, \ldots$, or any multiple of 180 degrees. In radians, $\theta \neq 0, \pi, 2\pi, 3\pi, \ldots$, or any multiple of π.

Ranges of tangent and cotangent

The ranges of both tangent and cotangent are infinite, which, when expressed in mathematical notation, looks like this: $-\infty < m(\theta) < \infty$ and $-\infty < n(\theta) < \infty$.

The range values for these functions get very small (toward negative infinity) or very large (toward positive infinity) whenever the denominator of the respective ratio gets close to 0. When you divide some number by a very small value, such as 0.0001, the result is large. The smaller the denominator, the larger the result.

Chapter 10

Applying Yourself to Trig Functions

*B*ack when trig functions were first developed or recognized — way back when — the motivation for creating the functions wasn't so men could sit around and say, "Hey, Caesar, did you know that the sine of 45 degrees is $\frac{\sqrt{2}}{2}$?"

Instead, the math gurus of the past worked out the principles of trigonometry because they needed some order or consistency to the numbers that they were applying to astronomy, agriculture, and architecture. They figured out the relationships among all these numbers and shared them with the rest of the known, civilized world.

First Things First: Elevating and Depressing

Mathematical problems that require the use of trig functions often have one of two related angles: the angle of elevation or the angle of depression. The scenarios that use these angles usually involve calculating distances that can't be physically measured. For example, these angles are used when finding the distance from an airplane to a point on the ground or the distance up to a balloon or another object above you. Use the trig functions to solve for the missing part of the ratio or the side of the imaginary right triangle.

Ptolemy: Part right and part wrong

Ptolemy, also known as Claudius Ptolemaeus, was a Greek citizen who lived in Alexandria, Egypt, from about A.D. 87 to 150. He was an astronomer, mathematician, and geographer. Ptolemy believed that the sun and other planets revolved around earth. At that time, only five planets were known, and he believed that they revolved around earth in this order: Mercury, Venus, the sun, Mars, Jupiter, and Saturn. This theory was known as the Ptolemaic system. It predicted the positions of the planets with reasonable accuracy, considering that they were naked-eye observations, not enhanced with telescopes. Ptolemy may not have really believed in this system of the planets, though — perhaps he used it only as a method of calculating their relative positions. This man did get it right, though, in determining that earth is a sphere, not flat — this theory affected much of his important work in geography and cartography. His works, including an error that had Asia extending too far to the east, probably influenced Columbus's decision to sail west for the Indies.

An *angle of elevation* is measured from the horizontal going upward. The horizontal is usually the ground, street, floor, or any other flat object. Even though the ground isn't perfectly flat or horizontal, you determine the measurements with the assumption that it is. In trig, you have to consider the optimal situation — focus on the big picture, rather than on the imperfections. Figure 10-1 shows an angle of elevation.

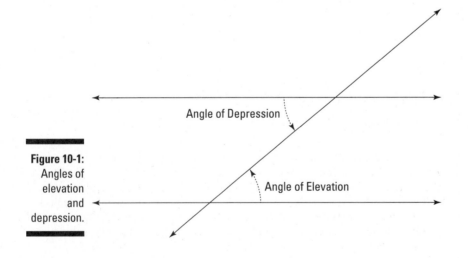

Figure 10-1:
Angles of
elevation
and
depression.

An *angle of depression* is measured from the horizontal going downward. For these angles, the horizontal is, for example, an airplane's flight path or a person's line of sight while standing on a mountaintop. The angle is formed when a person on that plane or mountaintop looks at an object on the ground (or on a path parallel to the ground). Refer to Figure 10-1 for a sample angle of depression.

Measuring Tall Buildings with a Single Bound

Every day, people use trigonometry to measure things that they can't reach. How high is that building? Will this ladder reach to the top of that tree? By using the appropriate trig functions, you can find answers to such questions. Two major considerations to keep in mind when working out problems by using trig are as follows: Which trig function should you use, and what are the units or measures in the answer?

The missing values in the ratios of the trig functions represent the missing parts in the problems. You assign the known values appropriately and solve for what's left.

Rescuing a damsel from a tower

Consider the oh-so-common scenario: A damsel is in distress and is being held captive in a tower. Her knight in shining armor is on the ground below with a ladder. He needs to know whether it'll reach her or whether he needs a longer ladder.

When the stunning knight stands 15 feet from the base of the tower and looks up at his precious damsel, the angle of elevation to her window is 60 degrees. How long does the ladder have to be? Figure 10-2 shows the situation in pictorial form.

1. **Identify the parts of the right triangle that you can use to solve the problem.**

   You know that the acute angle is 60 degrees, and the adjacent side of the triangle is along the ground; the distance from the vertex of the angle (where the knight is standing) to the base of the tower is 15 feet (the adjacent side). The hypotenuse is the length needed for the ladder — call it x. Figure 10-3 shows you the triangle.

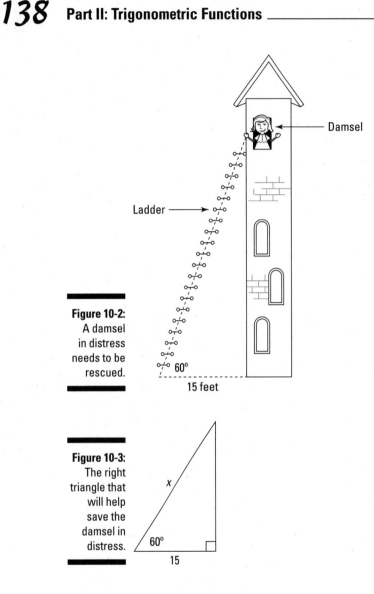

Figure 10-2: A damsel in distress needs to be rescued.

Figure 10-3: The right triangle that will help save the damsel in distress.

2. **Determine which trig function to use.**

   The adjacent side and hypotenuse are parts of the cosine ratio. Those sides are also parts of the secant ratio, but if at all possible, you should use the three main functions, not their reciprocals.

3. **Write an equation with the trig function; then insert the values that you know.**

   For a refresher on those values, look at the charts in Chapter 7. The cosine of 60 degrees is $\frac{1}{2}$, the adjacent side is 15 feet, and the hypotenuse is unknown.

$$\cos 60 = \frac{\text{adj}}{\text{hyp}}$$

$$\frac{1}{2} = \frac{15}{x}$$

4. Solve the equation.

Cross-multiplying, you get

$$\frac{1}{2} = \frac{15}{x}$$
$$1 \cdot x = 15 \cdot 2$$
$$x = 30$$

The ladder needs to be 30 feet long. (That knight had better be pretty strong!)

Determining the height of a tree

Suppose you're flying a kite, and it gets caught at the top of a tree. You've let out all 100 feet of string for the kite, and the angle that the string makes with the ground (the angle of elevation) is 75 degrees. Instead of worrying about how to get your kite back, you wonder, "How tall is that tree?"

Figure 10-4 shows the scenario.

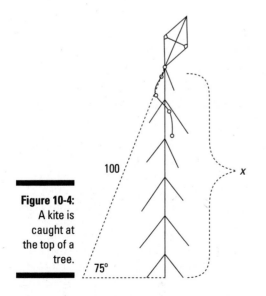

Figure 10-4:
A kite is caught at the top of a tree.

100

x

75°

To find a solution to your quandary, follow these steps:

1. **Identify the parts of the right triangle that you can use to solve the problem.**

   The hypotenuse of the right triangle is the length of the string. The side opposite the 75-degree angle is what you're solving for; call it x.

2. **Determine which trig function to use.**

   The hypotenuse and opposite side are part of the sine ratio.

3. **Write an equation with the trig function; then insert the values that you know.**

   The 75-degree angle isn't one of the more-common angles, so use a scientific calculator or one of the tables in the Appendix to obtain a value for the sine, correct to three decimal places. The sine of 75 degrees is about 0.966, the hypotenuse is 100 feet, and the opposite side is what is unknown.

   $$\sin 75 = \frac{\text{opp}}{\text{hyp}}$$
   $$0.966 = \frac{x}{100}$$

4. **Solve the equation.**

   Cross-multiplying, you get

   $$0.966 = \frac{x}{100}$$
   $$0.966 \cdot 100 = x$$
   $$96.6 = x$$

   The tree is over 96 feet tall. Lots of luck retrieving the kite.

Measuring the distance between buildings

Jumping Jehoshaphat makes his living by jumping, on his motorcycle, from building to building, cliff to bluff, or any place he can get attention for doing it. His record jump is a distance of 260 feet, from one building to another. Jehoshaphat is on to his next feat and needs to determine the distance from one building to another. His assistant, Lovely Lindsay, holds a 6-foot pole perpendicular to the roof she's standing on. When Jehoshaphat, standing on top of the first building, sights straight across to a point at the base of the pole and then sights a point halfway up the pole, the angle of elevation is 1 degree. Will he be able to make the jump? See Figure 10-5 for a visual of Jehoshaphat's calculation.

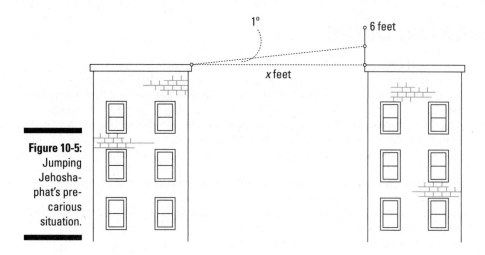

1. **Identify the parts of the right triangle that you can use to solve the problem.**

   You know the length of the side opposite the 1-degree angle, which is half the pole length (half is 3 feet), and the adjacent side is the unknown distance. Call that distance x.

2. **Determine which trig function to use.**

   The tangent of an angle uses opposite divided by adjacent.

3. **Write an equation with the trig function; then insert the values that you know.**

   The length of half the pole is 3 feet, so the equation looks like this:

   $$\tan 1° = \frac{\text{opp}}{\text{adj}} = \frac{3}{x}$$

4. **Solve for the value of x.**

   Use the Appendix or a calculator to find the value of the tangent of 1 degree.

   $$\tan 1° = \frac{3}{x}$$
   $$0.0175 = \frac{3}{x}$$
   $$x = \frac{3}{0.0175}$$
   $$= 171.4$$

   You find that the distance between the buildings is a little less than 172 feet across. Jehoshaphat should be able to make the jump easily, because his record is 260 feet.

Measuring Slope

Have you ever noticed a worker along the road, peering through an instrument, looking at a fellow worker holding up a sign or a flag? Haven't you ever wondered what they're doing? Have you wanted to get out and look through the instrument, too? With trigonometry, you can do just what those workers do — measure distances and angles. Land surveyors use trigonometry and their fancy equipment to measure things like the slope of a piece of land (how far it drops over a certain distance).

If you read the first section in this chapter, you may recognize that the slope of land downward is sort of like an angle of depression. Slopes, angles of depression, and angles of elevation are all interrelated because they use the same trig functions. It's just that in slope applications, you're solving for the angle rather than a length or distance.

To solve one of these surveying problems involving slope, you can use the trig ratios and right triangles. One side of the triangle is the distance from one worker to the other; the other side is the vertical distance from the ground to a point on a pole. You form a ratio with those measures and determine the angle — *voilà!*

Suppose that Elliott and Fred are making measurements for the road-paving crew. They need to know how much the land slopes downward along a particular stretch of road to be sure there's proper drainage. Elliott walks 80 feet from Fred and holds up a long pole, perpendicular to the ground, that has markings every inch along it. Fred looks at the pole through a sighting instrument. Looking straight across, parallel to the horizon, Fred sights a point on the pole 50 inches above the ground — call it point A. Then Fred looks through the instrument at the bottom of the pole, creating an angle of depression. See Figure 10-6 for a diagram of this situation. What is the angle of depression, or slope of the road, to where Elliott is standing?

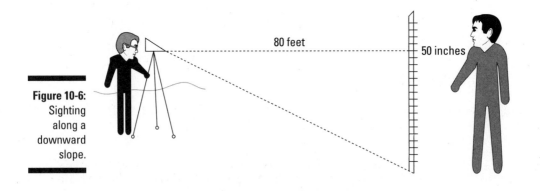

Figure 10-6:
Sighting
along a
downward
slope.

80 feet

50 inches

1. **Identify the parts of the right triangle that you can use to solve the problem.**

   The values you know are for the sides adjacent to and opposite the angle of depression. Call the angle measure x.

2. **Determine which trig function to use.**

   The tangent of the angle with measure x uses opposite divided by adjacent.

3. **Write an equation with the trig function; then insert the values that you know.**

   $$\tan x = \frac{\text{opposite}}{\text{adjacent}}$$

   In this problem, you need to write the equation with a common unit of measurement — either feet or inches. Changing 80 feet to inches makes for a big number; changing 50 inches to feet involves a fraction or decimal. Whichever unit you choose is up to you. In this example, I choose the big number, so I convert feet to inches.

   80 feet = 80 · 12 inches = 960 inches

   Substituting in the values, you get the tangent of some angle with a measure of x degrees:

   $$\tan x = \frac{\text{opp}}{\text{adj}} = \frac{50}{960} = 0.05208333$$

4. **Solve for the value of x.**

   In the Appendix, you see that an angle of 2.9 degrees has a tangent of 0.0507, and a 3-degree angle has a tangent of 0.0524. The 3-degree angle has a tangent that's closer to 0.05208333, so you can estimate that the road slopes at a 3-degree angle between Elliott and Fred.

Another way to solve for that angle measure is to use a scientific calculator and the inverse tangent function. I explain about inverse functions in Chapter 15, but you can jump ahead here and take advantage of technology. My calculator says that the angle whose tangent is 0.05208333 is an angle of 2.98146 degrees. So, the estimate of three degrees from the table is right on.

The Sky's (Not) the Limit

Early trigonometry had many earthbound applications — surveyors and engineers have used it for centuries. Over time, astronomers and navigators on journeys around the world began using trig to solve many mysteries here on earth and in outer space. They estimated or measured angles by sighting objects in the heavens and charting their movements. Then they used the angle between one sighting and another to solve for the distances that are unreachable.

Spotting a balloon

Cindy and Mindy, standing a mile apart, spot a hot-air balloon directly above a particular point on the ground somewhere between them. The angle of elevation from Cindy to the balloon is 60 degrees; the angle of elevation from Mindy to the balloon is 70 degrees. Figure 10-7 shows a visual representation. How high is the balloon?

If you look at Figure 10-7, you see that two right triangles are formed. The two triangles share a side — the one opposite the measured acute angle in each. Call the length of that shared side y. The two adjacent sides add up to 1 mile, so you can keep the variables to a minimum by naming one side x and the other $1 - x$. Figure 10-8 shows the triangles with the variables.

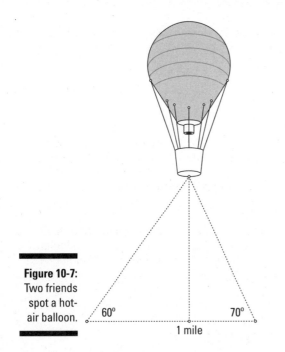

Figure 10-7:
Two friends
spot a hot-
air balloon.

60°   70°

1 mile

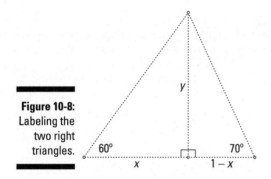

Figure 10-8:
Labeling the
two right
triangles.

To figure out how high the balloon is, follow these steps:

1. **Identify the parts of the triangles that you can use to solve the problem.**

   In both triangles, you have variables for the adjacent and opposite sides of the acute angles of elevation.

2. **Determine which trig function to use.**

   The tangent ratio uses the opposite and adjacent sides.

3. **Write equations with the trig functions.**

$$\tan 60° = \frac{\text{opp}}{\text{adj}} = \frac{y}{x} \text{ and } \tan 70° = \frac{\text{opp}}{\text{adj}} = \frac{y}{1-x}$$

$$\tan 60° = \frac{y}{x} \qquad \tan 70° = \frac{y}{1-x}$$

4. **Solve for *x* by setting the equations equal to one another.**

   First solve each of the equations for *y*.

$$\tan 60° = \frac{y}{x} \text{ and } \tan 70° = \frac{y}{1-x}$$

$$x \tan 60° = y \qquad (1-x) \tan 70° = y$$

   Set those two equations equal to one another and solve for *x*.

$$x \cdot \tan 60° = (1-x)\tan 70°$$

$$x \cdot \tan 60° = \tan 70° - x \tan 70°$$

$$x \cdot \tan 60° + x \cdot \tan 70° = \tan 70°$$

$$x (\tan 60° + \tan 70°) = \tan 70°$$

$$x = \frac{\tan 70°}{\tan 60° + \tan 70°}$$

5. Solve for the value of x.

You find the value of x by finding the values of the functions with a calculator or in the Appendix. Upon doing so, you find that x is approximately 0.613 miles. Put that value into one of the equations to solve for y:

$$x \cdot \tan 60° = y$$
$$(0.613)\tan 60° = y$$
$$(0.613)(1.732) = y$$
$$1.062 = y$$

The balloon is 1.062 miles high — sounds a tad high to me!

Tracking a rocket

In this example, a rocket is shot off and travels vertically as a scientist, who's a mile away, watches its flight. One second into the flight, the angle of elevation of the rocket is 30 degrees. Two seconds later, the angle of elevation is 60 degrees. How far did the rocket travel in those two seconds? Figure 10-9 shows the rocket rising vertically.

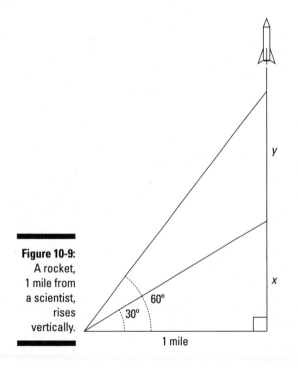

Figure 10-9: A rocket, 1 mile from a scientist, rises vertically.

1. **Identify the parts of the triangles that you can use to solve the problem.**

   In Figure 10-9, you see two right triangles. One is superimposed on the other and shares a side — the adjacent side. In both triangles, the relevant sides are those that are adjacent and opposite the angles of elevation.

2. **Determine which trig function to use.**

   The ratio of the tangent uses the adjacent and opposite sides.

3. **Write equations with the trig functions.**

   $$\tan 30° = \frac{\text{opp}}{\text{adj}} = \frac{x}{1} = x$$

   $$\tan 60° = \frac{\text{opp}}{\text{adj}} = \frac{x+y}{1} = x+y$$

4. **Solve for the values of *x* and *y*.**

   The tangents of 30-degree and 60-degree angles are convenient values. If you refer to the Appendix, you see that $\tan 30° = \frac{\sqrt{3}}{3} = x$ and $\tan 60° = \sqrt{3} = x + y$.

Sine and cosine with algebra

You can approximate, fairly accurately, the sine and cosine of angles with an *infinite series,* which is the sum of the terms of some sequence, or list, of numbers. Take note, however, that the series for sine and cosine are accurate only for angles from about –90 degrees to 90 degrees. The series for the sine of an angle is

$$\sin x = x - \frac{x^3}{3!} + \frac{x^5}{5!} - \frac{x^7}{7!} + \frac{x^9}{9!} - \dots \text{ and the}$$

series for the cosine of an angle is

$$\cos x = 1 - \frac{x^2}{2!} + \frac{x^4}{4!} - \frac{x^6}{6!} + \frac{x^8}{8!} - \dots. \text{ To use}$$

these formulas, you have to write the angle measure, *x,* in radians and carry out the computations several places. The exclamation points in the formulas don't mean "Oh, goodness! It's a 3!" The exclamation points are mathematical operations called *factorials. Factorial* means to multiply that number times every positive integer smaller than it. Going back to the series for the sine, an angle of 30 degrees is about 0.5236 radians. To find sin 0.5236, use the formula to get

$$\sin 0.5236 = 0.5236 - \frac{(0.5236)^3}{3!} + \frac{(0.5236)^5}{5!} - \frac{(0.5236)^7}{7!}$$

$$= 0.5236 - 0.0239 + 0.000328 - 0.000002$$

$$= 0.500026$$

The result is pretty close to the sine of 30 degrees, which is $\frac{1}{2}$. Carrying out the computations using a few more terms will make the result even closer to the actual answer. And the closer the angle measure is to 0, the more quickly the value of the sine or the cosine meets the exact value (the fewer terms are necessary for the answer).

The value of y is the distance that the rocket traveled between the first and second sightings, so, solving for y, you get

$$y = x + y - x = \sqrt{3} - \frac{\sqrt{3}}{3} = \frac{3\sqrt{3}}{3} - \frac{\sqrt{3}}{3} = \frac{2\sqrt{3}}{3} \approx 1.155.$$

The rocket rose about 1.155 miles in two seconds.

Measuring the view of satellite cameras

Consider a satellite that orbits earth at an altitude of 750 miles. Earth has a radius of 3,950 miles. How far in any direction can the satellite's cameras see? Figure 10-10 shows the satellite and the length of the camera's scope due to the curvature of earth.

   1. Identify the parts of the triangle that you can use to solve the problem.

      Because a satellite's line of sight is tangent to the curvature of earth, and tangents to a circle form 90-degree angles with radii of the circle, you can see two right triangles in Figure 10-10. The two sides of angle θ are the radius touching the tangent to the circle and the segment extending from the center of the circle up to the satellite. These sides are the hypotenuse and adjacent side of the right triangle with acute angle θ.

   2. Determine which trig function to use.

      The adjacent side and hypotenuse are part of the ratio for the cosine of θ.

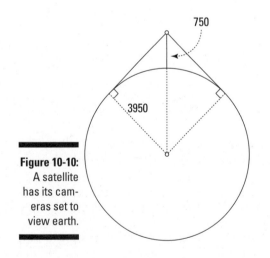

Figure 10-10:
A satellite has its cameras set to view earth.

3. **Write the equation with the trig function; then input the measures that you know and solve for cos θ.**

    The adjacent side measures 3,950 miles, and the hypotenuse is the sum of the radius and height of the satellite: 3,950 + 750 = 4,700 miles.

    $$\cos\theta = \frac{\text{adjacent}}{\text{hypotenuse}} = \frac{3,950}{4,700} \approx 0.8404$$

4. **Determine the value of θ.**

    Refer to the Appendix to find the angle whose cosine is closest to 0.8404. To the nearest degree, an angle of 33 degrees has this cosine.

5. **Determine how much of earth's circumference is covered in either direction from the satellite.**

    The satellite's line of sight goes 33 degrees in either direction, or 66 degrees total, which is $\frac{66}{360}$ of the entire circumference (because all the way around would be 360 degrees). If the radius of earth is 3,950 miles, then you can substitute that number into the equation for a circle's circumference: $C = 2\pi r = 2(3.14)(3,950) \approx 24,819$. That's the earth's circumference. The distance that the satellite scans, then, is $\frac{66}{360} \cdot 24,819 \approx 4,550$, or about 4,550 total miles or 2,275 miles in any direction.

Calculating Odd Shapes and Maneuvering Corners

Sometimes, finding a measure isn't so easy. You may have to deal with an irregular shape or even calculate your way around a fixed object. Whatever the case, you can use trigonometry to find the answers you've been searching for.

Finding the area of a triangular piece of land

The most commonly used formula for the area of a triangle is $A = \frac{1}{2}bh$, where A is the area, b is the length of the triangle's base, and h is the height of the triangle drawn perpendicular to that base. Figure 10-11 illustrates the different components of this formula.

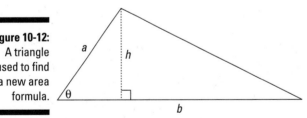

Figure 10-11:
A triangle's
base and
height.

This area formula works fine if you can get the measure of the base and the height, and if you can be sure that you've measured a height that's perpendicular to the side of the triangle. But what if you have a triangular yard — a *big* triangular yard — and have no way of measuring some perpendicular segment to one of the sides? One alternative is to use Heron's Formula, which uses the measures of all three sides. (I show you that one in the next section.) The other alternative, of course, is to use trigonometry — or, at least, a formula with an angle measure in it. To measure that angle, you can be very sophisticated and get a surveying apparatus, or if you've got a protractor handy, you can do a decent estimate by extending the sides at an angle for a bit and eyeballing the angle size.

The trig formula for finding the area of a triangle is $A = \frac{1}{2}ab\sin\theta$, where a and b are two sides of the triangle and θ is the angle formed between those two sides. You don't need the measure of the third side at all, and you certainly don't need a perpendicular side.

Using trigonometry, I show you where this formula comes from. Take a look at the triangle in Figure 10-12, with sides a and b and the angle between them.

Figure 10-12:
A triangle
used to find
a new area
formula.

Start with the traditional formula for the area of this triangle, $A = \frac{1}{2}bh$. Then look at the smaller triangle to the left. (Because the height is drawn perpendicular to the base, the sides and height form a right triangle.) The acute angle θ has a sine equivalent to the following: $\sin\theta = \dfrac{\text{opposite}}{\text{hypotenuse}} = \dfrac{h}{a}$. If you solve that equation for h by multiplying each side by a, you get

$$\sin\theta = \frac{h}{a}$$

$$a\sin\theta = h$$

Replace the h in the traditional formula with its equivalent from the equation above, and you get

$$A = \frac{1}{2}bh = \frac{1}{2}b(a\sin\theta) = \frac{1}{2}ab\sin\theta$$

Check out how this formula works in an actual problem. The triangle in Figure 10-13 shows the measures of two of its sides and the angle between them.

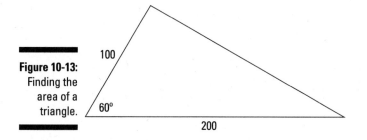

Figure 10-13:
Finding the area of a triangle.

To find the area of the triangle in Figure 10-13:

1. **Use the formula $A = \frac{1}{2}ab\sin\theta$, inserting the values that you know.**

   $$A = \frac{1}{2}(100)(200)\sin 60°$$

2. **Solve for the value of the area.**

   $$= \frac{1}{2}(100)(200) \cdot \frac{\sqrt{3}}{2}$$

   $$= 10,000 \cdot \frac{\sqrt{3}}{2} \approx 8,660$$

The area is about 8,660 square units.

Using Heron's Formula

As promised, in this section, I show you how to find the area of a triangle using Heron's Formula. Heron's Formula is especially helpful when you have access to the measures of the three sides of a triangle but can't draw a perpendicular height or don't have a protractor for measuring an angle.

Consider the situation where you have a large ball of string that's 100 yards long and you're told to mark off a triangular area — with the string as the marker for the border of the area. You walk 40 yards in one direction, take a turn, and walk another 25 yards; then you head back to where you started and use up that last 35 yards of string. How large an area have you created?

Heron's Formula reads: $A = \sqrt{s(s-a)(s-b)(s-c)}$ where a, b, and c are the lengths of the sides of the triangle and s is the semi-perimeter (half the perimeter).

In the case of your triangle and the string, the perimeter is $40 + 25 + 35 = 100$ yards. Half that is 50, so the formula now reads:

$$A = \sqrt{50(50-40)(50-25)(50-35)}$$
$$= \sqrt{50(10)(25)(15)} = \sqrt{187,500} \approx 433$$

You've marked off an area of approximately 433 square yards.

Moving an object around a corner

Here's an application of trigonometry that you may very well be able to relate to: Have you ever tried to get a large piece of furniture around a corner in a house? You twist and turn and put it up on end, but to no avail. In this example, pretend that you're trying to get a 15-foot ladder around a corner where two 4-foot-wide hallways meet at a 90-degree angle. Figure 10-14 shows a picture of the situation.

The tightest part comes when the ladder is halfway through the hallway, or when the angles where it touches the outer walls are the same. When the ladder is at the tightest point, it'll form a right triangle with equal sides — half the ladder to each side of the corner. Because the sides of the right triangle are equal at this point, you've got an isosceles right triangle, which has two 45-degree angles (see Figure 10-15, which shows the longest a ladder can be to fit around the corner). How long are the sides of the right triangle, then? When you know the dimensions of this isosceles right triangle, you can look at the hypotenuse — the ladder — and determine if it's short enough or too long to fit around the tightest part of this corner. And, of course, you don't want to scrape or punch holes in the wall!

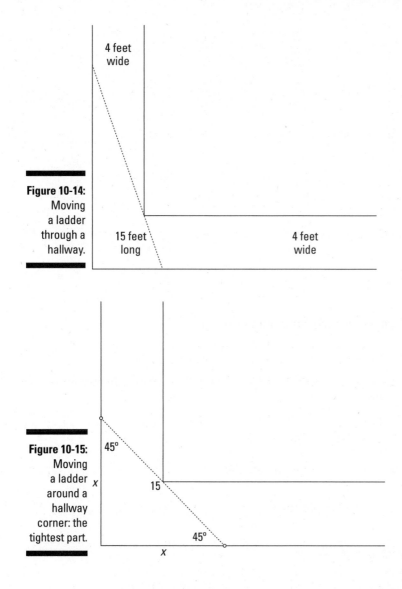

Figure 10-14:
Moving
a ladder
through a
hallway.

4 feet
wide

15 feet
long

4 feet
wide

Figure 10-15:
Moving
a ladder
around a
hallway
corner: the
tightest part.

45°

x

15

45°

x

1. **Determine the trig function that you can use with the measures available.**

   The hypotenuse is the length of the ladder — 15 feet. The opposite and
   adjacent sides are the same in an isosceles right triangle, and in this
   case, those two lengths are each 8 feet. You know this measure because
   all the triangles are isosceles right triangles, which means they have
   45-degree angles and equal leg measures (see Figure 10-16).

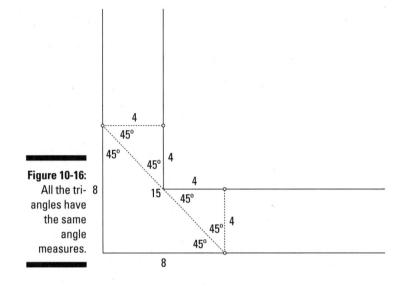

Figure 10-16:
All the tri-
angles have
the same
angle
measures.

2. **Determine which trig function to use.**

   Both sine and cosine include the length of the hypotenuse, which is what you're solving for, so you can use either function.

3. **Write the equation with the trig function; then insert the measures that you know.**

   $$\sin 45 = \frac{\text{opposite}}{\text{hypotenuse}} = \frac{8}{\text{hypotenuse}}$$

4. **Solve for the value of the hypotenuse.**

   $$\frac{\sqrt{2}}{2} = \frac{8}{\text{hypotenuse}}$$
   $$\sqrt{2} \cdot \text{hypotenuse} = 8 \cdot 2$$
   $$\text{hypotenuse} = \frac{16}{\sqrt{2}}$$
   $$\approx 11.314$$

   You find that at the tightest point around the corner, the hypotenuse is only slightly more than 11 feet. That 15-foot ladder will never fit around the corner.

Part III
Identities

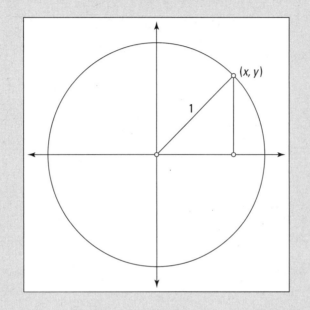

In this part...

- ✔ Develop a working list of the relationships between the trig functions.

- ✔ Use the trig identities to make statements more user-friendly.

- ✔ Determine the easiest way to solve a trig identity.

- ✔ Avoid going in endless loops with identities.

Chapter 11

Identifying Basic Identities

I'm sure a thousand questions are running through your mind: What's a trig identity? Is it possible to have a mistaken trig identity? Does anyone commit trig-identity theft? Can you have an identity crisis with trig? The answer: No — probably not.

Trig identities aren't nearly as sinister as you may think. They're actually very helpful tools in simplifying trig expressions and solving equations. These identities are special to trigonometry. Basically, they're equivalences — they give you options to substitute into equations in order to simplify. For example, wouldn't you rather use the number 1 than $\frac{1,623}{1,623}$? Of course! In most cases, the number 1 is simpler. That's how trig identities work — replace something with something simpler.

Identities are divided into different types, or categories, in order to help you remember them more easily and figure out when to use them more efficiently. In this chapter, I cover the gamut.

Flipping Functions on Their Backs: Reciprocal Identities

The simplest and most basic trig identities are those involving the reciprocals of the trig functions. To jog your memory, a *reciprocal* of a number is 1 divided by that number — for example, the reciprocal of 2 is $\frac{1}{2}$. Another way to describe reciprocals is to point out that the product of a number and its reciprocal is 1. In the case of 2 and its reciprocal, $\frac{1}{2}$, $2 \cdot \frac{1}{2} = 1$. The same principle goes for the trig reciprocals.

Here's how the reciprocal identities are defined:

- ✔ The reciprocal of sine is cosecant: $\frac{1}{\sin\theta} = \csc\theta$

- ✔ The reciprocal of cosine is secant: $\frac{1}{\cos\theta} = \sec\theta$

- ✔ The reciprocal of tangent is cotangent: $\frac{1}{\tan\theta} = \cot\theta$

- ✔ The reciprocal of cotangent is tangent: $\frac{1}{\cot\theta} = \tan\theta$

- ✔ The reciprocal of secant is cosine: $\frac{1}{\sec\theta} = \cos\theta$

- ✔ The reciprocal of cosecant is sine: $\frac{1}{\csc\theta} = \sin\theta$

In true fashion, when you multiply the reciprocals together, you get 1:

$$\sin\theta \cdot \csc\theta = 1$$

$$\cos\theta \cdot \sec\theta = 1$$

$$\tan\theta \cdot \cot\theta = 1$$

The reciprocal identity is a very useful one when you're solving trig equations — especially those involving fractions. If you find a way to multiply each side of an

A woman ahead of her time: Hypatia

One of the earliest recognized female mathematicians was Hypatia, who lived in Alexandria, Egypt, and is thought to have been born around A.D. 370. People of her day considered Hypatia to be not only a mathematician, but also a scientist and philosopher. Her father, Theon, was a professor of mathematics at the University of Alexandria. He taught Hypatia himself and shared with her his passion for knowledge and the search for answers. Hypatia developed a great enthusiasm for mathematics, as well as astronomy and astrology. Her father also believed in a strong and healthy body as well as mind, so he insisted on a regular physical routine to achieve this standard of excellence.

Hypatia is well known for her work on the ideas of *conic sections,* which are the curves that are formed by slicing a cone in various ways. The conic sections are called *parabola, circle, ellipse,* and *hyperbola.* She edited the work of Apollonius, making the concepts easier to understand. For this, she is considered to be the first woman to make a contribution resulting in the survival of some of the earlier mathematical ideas.

Hypatia came to a tragic end, killed by a mob that was spurred on by rumors created by leaders who didn't appreciate her religious stand or alliances.

equation by a function's reciprocal, you may be able to reduce some part of the equation to 1 — and simplifying is always a good thing.

Function to Function: Ratio Identities

Trig has two identities called *ratio identities*. This label can be confusing, because all the trig functions are defined by ratios. Somewhere along the line, however, mathematicians thought this description was perfect for these two identities, because they're basically fractions made up of two trig functions, one above the other, in each. The ratio identities create ways to write tangent and cotangent by using the other two basic functions, sine and cosine.

The ratio identities are $\tan\theta = \dfrac{\sin\theta}{\cos\theta}$ and $\cot\theta = \dfrac{\cos\theta}{\sin\theta}$.

These two identities come from the simplification of a couple of complex fractions. If you use the basic definitions for sine, cosine, and tangent —

$$\text{sine} = \frac{\text{opp}}{\text{hyp}}, \text{cosine} = \frac{\text{adj}}{\text{hyp}}, \text{and tangent} = \frac{\text{opp}}{\text{adj}}$$ — then you can use them to get

$$\frac{\text{sine}}{\text{cosine}} = \frac{\dfrac{\text{opp}}{\text{hyp}}}{\dfrac{\text{adj}}{\text{hyp}}} = \frac{\text{opp}}{\text{hyp}} \cdot \frac{\text{hyp}}{\text{adj}} = \frac{\text{opp}}{\text{adj}} = \text{tangent}$$

Likewise, because cotangent is the reciprocal of tangent,

$$\text{cotangent} = \frac{1}{\text{tangent}} = \frac{1}{\dfrac{\text{sine}}{\text{cosine}}} = \frac{\text{cosine}}{\text{sine}}$$

One little trick I've used over the years, to keep track of which ratio identity has sine over cosine and which is cosine over sine is to use the alphabet. Cotangent starts with the letter *c*, so I use this to determine that cosine over sine also starts with *c* (on the top). The ratio identity for tangent isn't quite as nice, but I know that tangent and sine are close together in the alphabet, so, since tangent starts with *t*, and *s* is pretty close to *t*, the ratio identity must start with *s*. Well, it works for me!

Opposites Attract: Opposite-Angle Identities

The *opposite-angle identities* change expressions with negative angles to equivalences with positive angles. Negative angles are great for describing a situation, but they aren't really handy when it comes to sticking them in a trig function and calculating that value. So, for example, you can rewrite the sine of –30 degrees as the sine of 30 degrees by putting a negative sign in front of the function:

$$\sin(-30°) = -\sin(30°)$$

This identity works for other angles, too. The angle measure doesn't have to be –30 degrees; any negative angle works. This negative-angle business works differently for different functions, though. First, consider the identities, and then see how they came to be.

The opposite-angle identities for the three most basic functions are

$$\sin(-\theta) = -\sin \theta$$

$$\cos(-\theta) = \cos \theta$$

$$\tan(-\theta) = -\tan \theta$$

The rule for the sine and tangent of a negative angle almost seems intuitive. But what's with the cosine? How can the cosine of a negative angle be the same as the cosine of the corresponding positive angle? Here's how it works.

If you refer back to Chapter 8, you find that the function values of angles with their terminal sides in the different quadrants have varying signs. Sine, for example, is positive when the angle's terminal side lies in the first and second quadrants, whereas cosine is positive in the first and fourth quadrants. In addition, Chapter 4 shows you how to draw angles on a coordinate plane: Positive angles go counterclockwise from the positive *x*-axis, and negative angles go clockwise.

With those points in mind, take a look at Figure 11-1, which shows a –45-degree angle and a 45-degree angle.

Figure 11-1:
Angles of
–45 degrees
and 45
degrees.

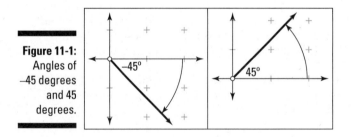

First, consider the –45-degree angle. This angle has its terminal side in the fourth quadrant, so its sine is negative. A 45-degree angle, on the other hand, has its terminal side in the first quadrant, so it has a positive sine. What about a negative angle whose terminal side ends up in the second quadrant, like –200 degrees? A +200-degree angle has a negative sine, because its terminal side is in the third quadrant, but a –200-degree angle has a positive sine, because its

---

A new way to slice pi-e

Two of the best-known symbols used for two constants in mathematics are the Greek letter pi, π, and the lowercase letter e. The value of π, approximately 3.14159, is a decimal that goes on forever. The value of e, approximately 2.71828, goes on forever, too.

The symbol for pi was first introduced in the early 1700s by William Jones, an obscure English writer who was composing a book for math beginners. Common belief is that he chose this letter because p is the first letter in *perimetron,* meaning perimeter. He didn't realize that what he was doing would have such a long-lasting effect on the world of mathematics. A few years later, the famous mathematician Leonhard Euler used the letter p instead of the Greek letter π, but he eventually adopted the Greek notation, too, making it even more popular with the rest of the world.

The letter e represents the base for natural logarithms. Like π, this value occurs naturally — in other words, the value e occurs as a multiplier or base of the equations that represent many natural phenomena. Euler also had a hand in popularizing this symbol sometime in the early to mid-1700s. The question comes to mind as to why he chose that particular letter. Using π or p seems natural for a concept linked to the perimeter of a circle, but the letter e isn't such an obvious choice. Here are a few possible reasons why our ancestors chose this letter: First, e is the first letter in *exponential,* which is closely tied to logarithms; or perhaps they used e because it's near the beginning of the alphabet and hadn't been used as a symbol for anything else. Could it be, instead, that Euler chose the letter e because it's the first letter in his name? Whatever the reason, it appears that π and e are here to stay.

---

terminal side is in the second quadrant. The values of the sines are opposites — further convincing you of the rule that the sine of a negative angle is the opposite value of that of the positive angle with the same measure.

Now on to the cosine function. In light of the cosine's sign with respect to the coordinate plane, you know that an angle of –45 degrees has a positive cosine. So does its counterpart, the angle of 45 degrees, which is why

$$\cos(-45°) = \cos(45°) = \frac{\sqrt{2}}{2}.$$

So, you see, the cosine of a negative angle is the same as that of the positive angle with the same measure.

Next, try the identity on another angle, a negative angle with its terminal side in the third quadrant. Figure 11-2 shows a negative angle with the measure of –120 degrees and its corresponding positive angle, 120 degrees.

The angle of –120 degrees has its terminal side in the third quadrant, so both its sine and cosine are negative. Its counterpart, the angle measuring 120 degrees, has its terminal side in the second quadrant, where the sine is positive and the cosine is negative. So, the sine of –120 degrees is the opposite of the sine of 120 degrees, and the cosine of –120 degrees is the same as the cosine of 120 degrees. In trig notation, it looks like this:

$$\sin(-120°) = -\sin(120°) = -\frac{\sqrt{3}}{2} \text{ and } \cos(-120°) = \cos(120°) = -\frac{1}{2}$$

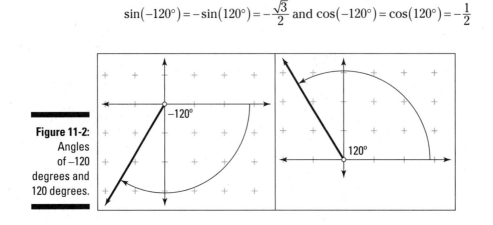

Figure 11-2:
Angles
of –120
degrees and
120 degrees.

When you apply the opposite-angle identity to the tangent of a 120-degree angle (which comes out to be negative), you get that the opposite of a negative is a positive. Surprise, surprise. So, applying the identity, the opposite makes the tangent positive, which is what you get when you take the tangent of 120 degrees, where the terminal side is in the third quadrant and is, therefore, positive.

But look at the opposite-angle identity for the tangent in another way: Use the ratio identity to prove why it works.

$$\tan(-\theta) = \frac{\sin(-\theta)}{\cos(-\theta)} = \frac{-\sin(\theta)}{\cos(\theta)} = -\frac{\sin(\theta)}{\cos(\theta)} = -\tan(\theta)$$

Revisiting the Classic Theorem: Pythagorean Identities

Good old Pythagoras is at work everywhere — his theorem keeps cropping up in the strangest places. (Not that a chapter in this book is really a strange place, of course.) This section takes you past the basics and expands on them with the three identities called *Pythagorean identities*. *(For more on Pythagoras's theorem, refer to Chapters 2 and 6.)*

The Pythagorean identities are building blocks for many of the manipulations of equations and expressions. They provide a generous number of methods for solving problems more efficiently, because they allow you to write complicated expressions in a much simpler form.

The Pythagorean identities are

$$\sin^2 \theta + \cos^2 \theta = 1$$

$$\tan^2 \theta + 1 = \sec^2 \theta$$

$$1 + \cot^2 \theta = \csc^2 \theta$$

The exponential notation used in these identities is peculiar to trigonometry. The expression $\sin^2 \theta$ actually means $(\sin \theta)^2$, which says, "Find the sine of angle θ and then square that number." But mathematicians hate to waste, and having to put those big, cumbersome parentheses around "$\sin \theta$" all the time seemed wasteful. So they agreed on a condensed version: The superscript 2 right after "sin" means that you square the whole expression. The same type of notation also goes for the other trig functions ($\cos^2 \theta$, $\tan^2 \theta$, $\cot^2 \theta$, and so on).

The mother of all Pythagorean identities

The Pythagorean identity that birthed the other two is $\sin^2 \theta + \cos^2 \theta = 1$. But, you may wonder, where did this identity come from, and why is it so important? Last things first: The primary Pythagorean identity is important because it sets a combination of functions equal to 1, and this simplification is very helpful for solving trig equations. As such, it's probably one of the most frequently used identities of them all. This identity comes from putting a right triangle inside the unit circle and substituting values and equations to come up with a whole new equation (see Figure 11-3).

In Chapter 9, you discover that in a circle, $\sin \theta = \frac{y}{r}$ and $\cos \theta = \frac{x}{r}$, where (x,y) are the coordinates of the point and r is the radius of the circle. The value of x is also the length of the adjacent side of the triangle (horizontal length), and y is the length of the opposite side (vertical length). In a unit circle, the radius is equal to 1. When you substitute the 1 for r in the equation, you find that $\sin \theta = \frac{y}{1} = y$ and $\cos \theta = \frac{x}{1} = x$. Hold that thought.

The Pythagorean theorem says that when you square the value of a right triangle's two legs and add the results together, you get the square of the hypotenuse. In mathematical notation, it looks like this: $a^2 + b^2 = c^2$. In the case of the right triangle on the unit circle, because the radius (which is also the hypotenuse) is 1, you can say that $x^2 + y^2 = 1^2$. Now replace the x with $\cos \theta$ and the y with $\sin \theta$, switch the two terms around, and you get $\sin^2 \theta + \cos^2 \theta = 1$.

If all the finagling just seems like a lot of hocus-pocus to you, check out this identity in action. Suppose the angle in question is 30 degrees. Using the values for the functions of a 30-degree angle (see the Appendix), $\sin 30° = \frac{1}{2}$ and $\cos 30° = \frac{\sqrt{3}}{2}$, and putting them into the identity, you get

$$\sin^2\theta + \cos^2\theta = 1$$

$$\left(\frac{1}{2}\right)^2 + \left(\frac{\sqrt{3}}{2}\right)^2 = \frac{1}{4} + \frac{3}{4} = 1$$

Voilà!

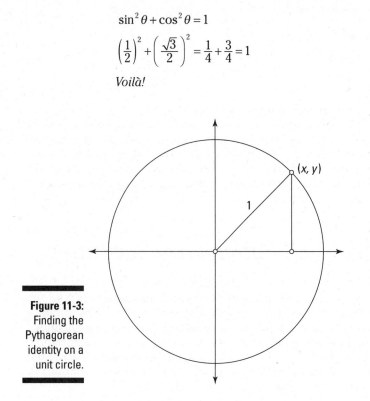

Figure 11-3:
Finding the
Pythagorean
identity on a
unit circle.

Extending to tangent and secant

The other two Pythagorean identities stem from the first one involving sine and cosine. All you do is throw in a little algebra, apply the reciprocal and ratio identities (see those sections earlier in this chapter), simplify, and — poof! — two new identities.

1. **Starting with the first Pythagorean identity, $\sin^2\theta + \cos^2\theta = 1$, divide each term by $\cos^2\theta$.**

$$\frac{\sin^2\theta}{\cos^2\theta} + \frac{\cos^2\theta}{\cos^2\theta} = \frac{1}{\cos^2\theta}$$

2. **Rewrite each term by using the exponential rule $\left(\frac{a^2}{b^2}\right) = \left(\frac{a}{b}\right)^2$.**

Note how the exponent 2 is pulled out of the parentheses:

$$\left(\frac{\sin\theta}{\cos\theta}\right)^2 + \left(\frac{\cos\theta}{\cos\theta}\right)^2 = \left(\frac{1}{\cos\theta}\right)^2$$

3. Replace each of the terms with an equivalent expression.

- Use a ratio identity, $\frac{\sin\theta}{\cos\theta} = \tan\theta$, to replace the first term with tan θ.
- Replace the second term with 1: $\frac{\cos\theta}{\cos\theta} = 1$.
- Use a reciprocal identity, $\frac{1}{\cos\theta} = \sec\theta$, to replace the third term with sec θ.

Substituting these expressions into the equation and simplifying, you get

$$(\tan\theta)^2 + (1)^2 = (\sec\theta)^2$$
$$\tan^2\theta + 1 = \sec^2\theta$$

This is the second Pythagorean identity.

Finishing up with cotangent and cosecant

If you read the preceding section, you can simply do a repeat performance with cotangent and cosecant. Here's how:

1. Starting with the first Pythagorean identity, $\sin^2\theta + \cos^2\theta = 1$, divide each term by $\sin^2\theta$.

$$\frac{\sin^2\theta}{\sin^2\theta} + \frac{\cos^2\theta}{\sin^2\theta} = \frac{1}{\sin^2\theta}$$

2. Rewrite each term by using the exponential rule $\left(\frac{a^2}{b^2}\right) = \left(\frac{a}{b}\right)^2$.

$$\left(\frac{\sin\theta}{\sin\theta}\right)^2 + \left(\frac{\cos\theta}{\sin\theta}\right)^2 = \left(\frac{1}{\sin\theta}\right)^2$$

3. Replace each of the terms with an equivalent expression.

- Replace the first term with 1: $\frac{\sin\theta}{\sin\theta} = 1$.
- Use a ratio identity, $\frac{\cos\theta}{\sin\theta} = \cot\theta$, to replace the second term with cot θ.
- Use a reciprocal identity, $\frac{1}{\sin\theta} = \csc\theta$, to replace the third term with csc θ.

Substituting these expressions into the equation and simplifying, you find that the result is

$$(1)^2 + (\cot\theta)^2 = (\csc\theta)^2$$
$$1 + \cot^2\theta = \csc^2\theta$$

and you have the last Pythagorean identity.

Rearranging the Pythagorean identities

The preceding sections show the original definitions of the Pythagorean identities, but as you probably suspected, the forms don't end there. Familiarizing yourself with the other versions of these identities is helpful so that you can easily recognize them when solving equations or simplifying expressions.

All these different versions have their places in trigonometric applications, calculus, or other math topics. You don't have to memorize them, because if you just remember the three basic Pythagorean identities, you can solve for what you need.

Changing $sin^2 \theta + cos^2 \theta = 1$

You can alter the original Pythagorean identity in myriad ways. For starters, you can isolate either $\sin^2 \theta$ or $\cos^2 \theta$ on one side of the equation by subtracting the other term:

$$\sin^2 \theta = 1 - \cos^2 \theta$$
$$\cos^2 \theta = 1 - \sin^2 \theta$$

Continuing on, you can factor the right side of either of these equations because that side is the difference of two perfect squares:

$$\sin^2 \theta = 1 - \cos^2 \theta = (1 - \cos \theta)(1 + \cos \theta)$$
$$\cos^2 \theta = 1 - \sin^2 \theta = (1 - \sin \theta)(1 + \sin \theta)$$

Sometimes, however, having an expression for $\sin \theta$ or $\cos \theta$, where the functions aren't squared, is helpful. Beginning with the earlier version of the basic Pythagorean identities, where one function is by itself, you can take the square root of each side to get

$$\sin\theta = \pm\sqrt{1 - \cos^2\theta} \ \text{ or } \cos\theta = \pm\sqrt{1 - \sin^2\theta}$$

Adjusting $tan^2 \theta + 1 = sec^2 \theta$

You can also adapt this second Pythagorean identity in various ways. Solving for $\tan^2 \theta$ by subtracting 1 from each side of the equation, you get

$$\tan^2 \theta = \sec^2 \theta - 1$$

Then, factoring the difference of the squares on the right (because that side is the difference of two perfect squares), you have

$$\tan^2 \theta = \sec^2 \theta - 1 = (\sec \theta - 1)(\sec \theta + 1)$$

John Napier, inventor of logs and bones

John Napier was a 16th-century Scottish mathematician and inventor credited with inventing logarithms (for which he is best known), the decimal point, and *Napier's Bones* — an early calculating instrument. Napier's Bones were actually strips of wood or bone with multiplication tables inscribed on them. People used them for multiplying, dividing, taking square and cube roots, determining decimal values of fractions, and doing computations with exponential and trig functions.

As an inventor, Napier created a hydraulic screw, which was used in coal pits to lower the water levels. He invented, or at least proposed, several military inventions as well. He proposed special artillery; bulletproof clothing; a submarine-like vehicle; and huge, burning mirrors to set enemy ships on fire.

Lastly, beginning with the earlier version and taking the square root of each side, you get

$$\tan\theta = \pm\sqrt{\sec^2\theta - 1}$$

Taking another approach to this Pythagorean identity, you can subtract $\tan^2\theta$ from each side and factor the result to get

$$1 = \sec^2\theta - \tan^2\theta = (\sec\theta - \tan\theta)(\sec\theta + \tan\theta)$$

The choice of restructuring always just depends on what you want to do with the terms — what other functions they'll be interacting with in a problem.

Reconfiguring $1 + \cot^2\theta = \csc^2\theta$

You can rearrange the last Pythagorean identity, too, by subtracting 1 from each side or by subtracting $\cot^2\theta$ from each side. The two new versions are

$$\cot^2\theta = \csc^2\theta - 1$$
$$1 = \csc^2\theta - \cot^2\theta$$

Each of the preceding equations has the difference of two perfect squares, which you can factor:

$$\cot^2\theta = \csc^2\theta - 1 = (\csc\theta - 1)(\csc\theta + 1)$$
$$1 = \csc^2\theta - \cot^2\theta = (\csc\theta - \cot\theta)(\csc\theta + \cot\theta)$$

And last, the square root of each side yields an identity involving just $\cot\theta$:

$$\cot\theta = \pm\sqrt{\csc^2\theta - 1}$$

Combining the Identities

Even though each function is perfectly wonderful, being able to express each of them in terms of all the other five trig functions is frequently to your advantage. For example, you may have an equation or expression with a lot of sines, but not all the terms are sines. Having them all match — all be in terms of sine — would help you solve the equation.

Armed with the reciprocal identities, ratio identities, and Pythagorean identities, you can do just that — write any trig function in terms of the others. In this section, I show you the many variations of sine; by applying some of the same identities and following similar steps, you can form a multitude of variations of the other trig functions. These five ways of writing sine in terms of the other five functions show you how powerful, versatile, and useful all these identities are. A word of warning: Some of these equations aren't very pretty. But, then beauty is in the eye of the beholder. And you may think these are just dandy.

The many faces of sine

Here are the five ways of expressing the sine function in terms of the other functions:

- ✔ Sine in terms of cosine: $\sin\theta = \pm\sqrt{1 - \cos^2\theta}$
- ✔ Sine in terms of tangent: $\sin\theta = \dfrac{\tan\theta}{\pm\sqrt{\tan^2\theta + 1}}$
- ✔ Sine in terms of cotangent: $\sin\theta = \dfrac{1}{\pm\sqrt{1 + \cot^2\theta}}$
- ✔ Sine in terms of secant: $\sin\theta = \dfrac{\pm\sqrt{\sec^2\theta - 1}}{\sec\theta}$
- ✔ Sine in terms of cosecant: $\sin\theta = \dfrac{1}{\csc\theta}$

Working out the versions

Choosing a good starting point helps — and makes for a nicer result. You don't want any expression that's too messy or hard to remember. Take advantage of identities that have your target function isolated in a single term. This section shows you the most typical methods for changing one trig function to another.

Changing sine to cosine

You can express the sine function in terms of cosine without doing much work.

1. **Starting with the Pythagorean identity involving sin θ and cos θ subtract cos² θ from each side.**

   $\sin^2\theta = 1 - \cos^2\theta$

2. **Take the square root of both sides.**

$$\sqrt{\sin^2 \theta} = \pm\sqrt{1 - \cos^2 \theta}$$

$$\sin\theta = \pm\sqrt{1 - \cos^2 \theta}$$

You can use either the positive root or the negative root, depending on your application.

Changing sine to tangent

To rewrite the sine function in terms of tangent:

1. **Start with the ratio identity involving sine, cosine, and tangent, and multiply each side by cosine to get the sine alone on the left.**

$$\cancel{\cos\theta} \cdot \frac{\sin\theta}{\cancel{\cos\theta}} = \tan\theta \cdot \cos\theta$$

$$\sin\theta = \tan\theta\cos\theta$$

2. **Replace cosine with its reciprocal function.**

$$\sin\theta = \tan\theta\left(\frac{1}{\sec\theta}\right)$$

3. **Solve the Pythagorean identity $\tan^2\theta + 1 = \sec^2\theta$ for secant.**

   This equation gives you $\pm\sqrt{\tan^2 \theta + 1} = \sec\theta$.

4. **Replace the secant in the sine equation (from Step 2).**

   You end up with $\sin\theta = \tan\theta\left(\dfrac{1}{\pm\sqrt{\tan^2 \theta + 1}}\right) = \dfrac{\tan\theta}{\pm\sqrt{\tan^2 \theta + 1}}$.

Not very pretty, but, if your equation has terms with tangents in them, you have a better chance of combining terms or reducing fractions.

Changing sine to cotangent

To write the sine function in terms of cotangent, begin with the equation you end up with in the preceding section,

$$\sin\theta = \tan\theta\left(\frac{1}{\pm\sqrt{\tan^2 \theta + 1}}\right) = \frac{\tan\theta}{\pm\sqrt{\tan^2 \theta + 1}}.$$

1. **Replace all the tangents with 1 over the reciprocal for tangent (which is cotangent) and simplify the expression.**

$$\sin\theta = \left(\frac{\dfrac{1}{\cot\theta}}{\pm\sqrt{\dfrac{1}{\cot^2\theta} + 1}}\right)$$

The result is a *complex fraction* — it has fractions in both the numerator and denominator — so it'll look a lot better if you simplify it.

2. **Rewrite the part under the radical as a single fraction and simplify it by using the law of exponents/radicals, taking the square root of each part.**

$$\sin\theta = \frac{\dfrac{1}{\cot\theta}}{\pm\sqrt{\dfrac{1}{\cot^2\theta}+\dfrac{\cot^2\theta}{\cot^2\theta}}} = \frac{\dfrac{1}{\cot\theta}}{\pm\sqrt{\dfrac{1+\cot^2\theta}{\cot^2\theta}}}$$

$$= \frac{\dfrac{1}{\cot\theta}}{\pm\dfrac{\sqrt{1+\cot^2\theta}}{\sqrt{\cot^2\theta}}} = \frac{\dfrac{1}{\cot\theta}}{\pm\dfrac{\sqrt{1+\cot^2\theta}}{\cot\theta}}$$

3. **Multiply the numerator by the reciprocal of the denominator.**

$$= \frac{\dfrac{1}{\cot\theta}}{\pm\dfrac{\sqrt{1+\cot^2\theta}}{\cot\theta}} = \frac{1}{\cancel{\cot\theta}}\cdot\frac{\cancel{\cot\theta}}{\pm\sqrt{1+\cot^2\theta}} = \frac{1}{\pm\sqrt{1+\cot^2\theta}}$$

That's it. And it even turned out simpler-looking than the sine written with tangents.

Changing sine to secant

The next function to define sine is the secant function. You see more radicals on the horizon, but radicals can be "tamed."

1. **Start with the sine in terms of the cosine (refer to the first change-up in this section).**

$$\sin\theta = \pm\sqrt{1-\cos^2\theta}$$

2. **Now replace the cosine with 1 over its reciprocal.**

$$\sin\theta = \pm\sqrt{1-\cos^2\theta}$$

$$= \pm\sqrt{1-\frac{1}{\sec^2\theta}}$$

The radical has a fraction in it. A better form is to simplify that fraction, so find a common denominator and split the fraction into two radicals — the bottom one of which you can further simplify:

$$\sin\theta = \pm\sqrt{\frac{\sec^2\theta}{\sec^2\theta}-\frac{1}{\sec^2\theta}} = \pm\sqrt{\frac{\sec^2\theta-1}{\sec^2\theta}} = \pm\frac{\sqrt{\sec^2\theta-1}}{\sec^2\theta}$$

Changing sine to cosecant

The last function to write sine in terms of is the cosecant — I saved the best (easiest) for last. The reciprocal of cosecant is sine, so this equation is just one of the basic reciprocal identities: $\sin\theta = \dfrac{1}{\csc\theta}$

Chapter 12

Operating on Identities

. . . .

. . . .

*T*he basic building-block identities are the reciprocal, ratio, and Pythagorean identities, which I discuss in detail in Chapter 11. In this chapter, you take those identities a step further and develop new identities, discovering how to add, subtract, multiply, and divide the trig functions — in particular, the nice values for angles of 0, 30, 45, 60, and 90 degrees. (Those angles aren't the only ones that you can perform operations on; they're just the most convenient to use when showing how the trig identities work.) By performing such operations, you can determine the function values of even more angles than before. Whole new worlds will open up to you!

Summing It Up

The sums of angles are covered by three basic identities; these identities involve sine, cosine, and tangent. After you recognize these three identities, you can adapt them for the other three functions (cosecant, secant, and cotangent) by using the reciprocal identities (detailed in Chapter 11). All you do is start with a basic sum identity, use a reciprocal identity to change the expression to the one you want, do the necessary simplifying, and then use the new sum identity as needed. You won't have to do this very often; you can usually get by with one of the three basics.

Use the angle-sum identities to find the function values of many, many angles, but the examples in this section just show the most convenient combinations — ones with exact values that you can fill into the formulas. Suppose, for example, that you want to find the exact value of the sine of

75 degrees. To minimize fuss, you can use the sum of 30 degrees and 45 degrees and the appropriate identity.

TRIG RULES
1
+1
2

The angle-sum identities find the function value for the sum of angle α and angle β:

$$\sin(\alpha + \beta) = \sin \alpha \cos \beta + \cos \alpha \sin \beta$$
$$\cos(\alpha + \beta) = \cos \alpha \cos \beta - \sin \alpha \sin \beta$$
$$\tan(\alpha + \beta) = \frac{\tan \alpha + \tan \beta}{1 - \tan \alpha \tan \beta}$$

Now for an example using a sum-of-angles identity.

Using the identity for the sine of a sum, find the sine of 75 degrees:

1. **Determine two angles whose sum is 75 for which you know the values for both sine and cosine.**

   Choose 30 + 45, not 50 + 25 or 70 + 5, because sticking to the more-common angles that have nice, exact values to use in the formula is your best bet.

2. **Input the angle measures into the identity.**

$$\sin(\alpha + \beta) = \sin \alpha \cos \beta + \cos \alpha \sin \beta$$
$$\sin(30 + 45) = \sin 30 \cos 45 + \cos 30 \sin 45$$

3. **Replace the functions of the angles with their values and simplify.**

$$\sin(30 + 45) = \sin 30 \cos 45 + \cos 30 \sin 45$$
$$= \frac{1}{2} \cdot \frac{\sqrt{2}}{2} + \frac{\sqrt{3}}{2} \cdot \frac{\sqrt{2}}{2}$$
$$= \frac{\sqrt{2}}{4} + \frac{\sqrt{6}}{4} = \frac{\sqrt{2} + \sqrt{6}}{4} = \sin(75)$$

Sometimes, you have more than one choice for the sum. In this next example, find the cosine of 120 degrees by using the identity for the cosine of a sum.

1. **Determine two angles whose sum is 120.**

   Choosing among the most convenient angles, you can use either 90 + 30 or 60 + 60. For this example, I use 90 + 30, because the sine of a 90-degree angle is 1, and the cosine is equal to 0. Both of those numbers are very nice to have in a computation because they keep it simple.

2. **Input the values into the identity.**

$$\cos(\alpha + \beta) = \cos \alpha \cos \beta - \sin \alpha \sin \beta$$
$$\cos(90 + 30) = \cos 90 \cos 30 - \sin 90 \sin 30$$

3. Replace the functions with their values and simplify.

$$\cos(90+30) = \cos 90 \cos 30 - \sin 90 \sin 30$$

$$= 0 \cdot \frac{\sqrt{3}}{2} - 1 \cdot \frac{1}{2} = 0 - \frac{1}{2} = -\frac{1}{2} = \cos(120)$$

But what if you want to use a different set of angles to find the cosine of 120 degrees?

1. Determine two angles whose sum is 120.

The only other combination that comes quickly to mind is to use $60 + 60$.

2. Input the values into the identity.

$$\cos(\alpha + \beta) = \cos \alpha \cos \beta - \sin \alpha \sin \beta$$

$$\cos(60+60) = \cos 60 \cos 60 - \sin 60 \sin 60$$

3. Replace the functions with their values and simplify.

$$\cos(60+60) = \cos 60 \cos 60 - \sin 60 \sin 60$$

$$= \frac{1}{2} \cdot \frac{1}{2} - \frac{\sqrt{3}}{2} \cdot \frac{\sqrt{3}}{2} = \frac{1}{4} - \frac{3}{4} = -\frac{2}{4} = -\frac{1}{2} = \cos(120)$$

It really doesn't matter which pair you use — you get the same answer.

These identities work with radian measures, too, such as finding $\tan \frac{7\pi}{12}$ by using the identity for the tangent of the sum of angles.

1. Determine two angles whose sum is $\tan \frac{7\pi}{12}$.

It may be easier to think of finding two numbers that add up to $\frac{7}{12}$, and leave the π off for a moment.

The two fractions that come to mind are $\frac{1}{3}$ and $\frac{1}{4}$. Because $\frac{1}{3} + \frac{1}{4} = \frac{4}{12} + \frac{3}{12} = \frac{7}{12}$ you have $\frac{7\pi}{12} = \frac{\pi}{3} + \frac{\pi}{4}$.

2. Input the values into the identity.

$$\tan(\alpha + \beta) = \frac{\tan \alpha + \tan \beta}{1 - \tan \alpha \tan \beta}$$

$$\tan\left(\frac{\pi}{3} + \frac{\pi}{4}\right) = \frac{\tan \frac{\pi}{3} + \tan \frac{\pi}{4}}{1 - \tan \frac{\pi}{3} \tan \frac{\pi}{4}}$$

3. Replace the functions with their values and simplify.

$$\tan\left(\frac{\pi}{3} + \frac{\pi}{4}\right) = \frac{\sqrt{3} + 1}{1 - \sqrt{3} \cdot 1} = \frac{\sqrt{3} + 1}{1 - \sqrt{3}} = \tan \frac{7\pi}{12}$$

The result in the last step doesn't leave the answer in the nicest form. The denominator has two terms, and one of them is a radical. One way to make the answer look a bit better and more intelligible is to use a technique called *rationalization.*

To rationalize the numerator or denominator of a fraction, multiply both the numerator and denominator by the *conjugate* (same terms, opposite sign) of the part that you're rationalizing. When you do so, you end up with the difference of two squares, which lets you get rid of the offending portion.

For the last example, you rationalize to get the radical out of the denominator:

$$\frac{\sqrt{3}+1}{1-\sqrt{3}} \cdot \frac{1+\sqrt{3}}{1+\sqrt{3}} = \frac{\sqrt{3}+3+1+\sqrt{3}}{1-\sqrt{3}} = \frac{4+2\sqrt{3}}{-2} = -2-\sqrt{3}$$

The final answer is a bit nicer to understand and estimate. Because $\sqrt{3}$ is about 1.7, you can estimate that $-2-\sqrt{3} \approx -2-1.7 = -3.7$.

Next, I come at these angle-sum identities from a different direction. Sometimes, you may not know what the angle measure is, but you know something about the angle's function values. For example, suppose you have two angles, α in QII and β in QI. You know that $\sin\alpha = \frac{3}{5}$ and $\cos\beta = \frac{24}{25}$. With that information, then what are $\sin(\alpha+\beta)$ and $\cos(\alpha+\beta)$?

1. **Find all the necessary function values for the sums.**

   Both the sine and cosine angle-sum identities use the sine and cosine of each angle involved. You already know the sine of one angle and the cosine of the other angle, so you have to determine the unknown cosine and sine — you can do so by using a Pythagorean identity:

   • First, use the value for sin α to solve for cos α:

   $$\sin^2\alpha + \cos^2\alpha = 1$$

   $$\left(\frac{3}{5}\right)^2 + \cos^2\alpha = 1$$

   $$\cos^2\alpha = 1 - \frac{9}{25} = \frac{16}{25}$$

   $$\cos\alpha = \pm\sqrt{\frac{16}{25}} = \pm\frac{4}{5}$$

   You end up with two results. Because the terminal side of angle α is in the second quadrant, the cosine of α, in this case, is negative:

   $$\cos\alpha = -\frac{4}{5}$$

• Now use the value for cos β to solve for sin β:

$$\sin^2\beta + \cos^2\beta = 1$$

$$\sin^2\beta + \left(\frac{24}{25}\right)^2 = 1$$

$$\sin^2\beta = 1 - \frac{576}{625} = \frac{49}{625}$$

$$\sin\beta = \pm\sqrt{\frac{49}{625}} = \pm\frac{7}{25}$$

Again, there are two different signs to choose from for the sin of β. The terminal side of angle β is in the first quadrant, where the sine is positive: $\sin\beta = \frac{7}{25}$.

2. **Insert the function values into the identities for the sine and cosine of the sum of angles.**

$$\sin(\alpha + \beta) = \sin\alpha\cos\beta + \cos\alpha\sin\beta$$

$$\sin(\alpha + \beta) = \left(\frac{3}{5}\right)\left(\frac{24}{25}\right) + \left(-\frac{4}{5}\right)\left(\frac{7}{25}\right)$$

$$\cos(\alpha + \beta) = \cos\alpha\cos\beta - \sin\alpha\sin\beta$$

$$\cos(\alpha + \beta) = \left(-\frac{4}{5}\right)\left(\frac{24}{25}\right) - \left(\frac{3}{5}\right)\left(\frac{7}{25}\right)$$

3. **Simplify the identities and solve for the answers.**

$$\sin(\alpha + \beta) = \left(\frac{3}{5}\right)\left(\frac{24}{25}\right) + \left(-\frac{4}{5}\right)\left(\frac{7}{25}\right)$$

$$= \left(\frac{72}{125}\right) + \left(-\frac{28}{125}\right) = \frac{44}{125}$$

$$\cos(\alpha + \beta) = \left(-\frac{4}{5}\right)\left(\frac{24}{25}\right) - \left(\frac{3}{5}\right)\left(\frac{7}{25}\right)$$

$$= \left(-\frac{96}{125}\right) - \left(\frac{21}{125}\right) = -\frac{117}{125}$$

By looking at the angle measures, you can predict whether the function value will be positive or negative. In the preceding example, the smaller angles, when added together, create an angle with its terminal side in the second quadrant. In Chapter 9, you find out that the sine of an angle in the second quadrant is positive. So, it's no surprise that the sine comes out to be a positive value and, likewise, that the cosine is a negative value (because cosine is negative in the second quadrant).

Overcoming the Differences

By adding angles together, you enlarge your repertoire. You have a longer list of *exact* function values — not just the basic function values, but also all the possible sums of these more-common angles. In like fashion, you have even more possibilities for finding the function values of angles when you use subtraction. For example, you can determine the sine of 15 degrees by using 45 degrees and 30 degrees and the appropriate identity.

The subtraction, or difference, identities find the function for the difference between angles α and β:

$$\sin(\alpha - \beta) = \sin \alpha \cos \beta - \cos \alpha \sin \beta$$

$$\cos(\alpha - \beta) = \cos \alpha \cos \beta + \sin \alpha \sin \beta$$

$$\tan(\alpha - \beta) = \frac{\tan \alpha - \tan \beta}{1 + \tan \alpha \tan \beta}$$

Notice how each of the subtraction identities resembles its corresponding angle-sum identity. For the sine rule, the sign between the two products changed from + to –, which seems to make sense. The opposite is true for cosine. The addition rule for cosine has – in it, and the subtraction (or difference) rule has + in it. The tangent rule has both + and – in it; the operation in the numerator mirrors the type of identity.

Only the original three trig functions have truly usable difference identities — the identities for the reciprocal functions are pretty darned complicated. If you want the difference of a reciprocal function, your best bet is to use the corresponding basic identity and find the reciprocal of the numerical answer after you're all finished.

To see one of the subtraction identities in action, check out the following example, which shows how you can find the sine of 15 degrees.

1. **Determine two angles with a difference of 15 degrees.**

   To keep things simple, use 45 and 30.

2. **Substitute the angles into the identity for the sine of a difference.**

   $$\sin(\alpha - \beta) = \sin \alpha \cos \beta - \cos \alpha \sin \beta$$
   $$\sin 15 = \sin(45 - 30) = \sin 45 \cos 30 - \cos 45 \sin 30$$

3. **Replace the terms with the function values and simplify the answer.**

   $$\sin(45 - 30) = \frac{\sqrt{2}}{2} \cdot \frac{\sqrt{3}}{2} - \frac{\sqrt{2}}{2} \cdot \frac{1}{2}$$
   $$= \frac{\sqrt{6}}{4} - \frac{\sqrt{2}}{4} = \frac{\sqrt{6} - \sqrt{2}}{4} = \sin(15)$$

Using radians introduces even more fractions into the picture, such as finding $\tan\frac{\pi}{12}$ by using the identity for the tangent of a difference.

1. **Determine which angles you need to get the difference.**

   The two angles are $\frac{\pi}{3}$ and $\frac{\pi}{4}$, giving you $\frac{\pi}{3}-\frac{\pi}{4}=\frac{4\pi}{12}-\frac{3\pi}{12}=\frac{\pi}{12}$.

2. **Substitute the angles into the identity for the tangent of a difference.**

$$\tan(\alpha-\beta)=\frac{\tan\alpha-\tan\beta}{1+\tan\alpha\tan\beta}$$

$$\tan\left(\frac{\pi}{12}\right)=\tan\left(\frac{\pi}{3}-\frac{\pi}{4}\right)=\frac{\tan\frac{\pi}{3}-\tan\frac{\pi}{4}}{1+\tan\frac{\pi}{3}\tan\frac{\pi}{4}}$$

3. **Replace the terms with the function values and simplify the answer.**

$$\tan\left(\frac{\pi}{3}-\frac{\pi}{4}\right)=\frac{\sqrt{3}-1}{1+\sqrt{3}\cdot 1}=\frac{\sqrt{3}-1}{1+\sqrt{3}}=\tan\left(\frac{\pi}{12}\right)$$

The result is rather messy. You can simplify it even more by multiplying the numerator and denominator by the *conjugate* (same terms, different sign) of the denominator and simplifying the result:

$$\frac{\sqrt{3}-1}{1+\sqrt{3}}\cdot\frac{1-\sqrt{3}}{1-\sqrt{3}}=\frac{\sqrt{3}-3-1+\sqrt{3}}{1-3}=\frac{2\sqrt{3}-4}{-2}=-\sqrt{3}+2$$

In Chapter 11, I explain the opposite-angle identities. This next example uses the identity for the cosine of a difference along with the angle measuring 0 degrees to create an opposite-angle identity. You may like this explanation better than those in Chapter 11, and it just goes to show you how versatile and user-friendly trig identities are — and how they all get along so well together.

In this example, find $\cos\left(-\frac{\pi}{3}\right)$ by using the identity for the difference between angles.

1. **Determine which angles you need to get the difference.**

   Using 0 and $\frac{\pi}{3}$ and subtracting with the 0 first gives a negative result:
   $0-\frac{\pi}{3}=-\frac{\pi}{3}$.

2. **Substitute the angles into the identity for the cosine of a difference.**

$$\cos(\alpha-\beta)=\cos\alpha\cos\beta+\sin\alpha\sin\beta$$

$$\cos\left(0-\frac{\pi}{3}\right)=\cos 0\cos\frac{\pi}{3}+\sin 0\sin\frac{\pi}{3}$$

3. Replace the angles with the function values and simplify the answer.

$$\cos\left(-\frac{\pi}{3}\right) = \cos 0 \cos\frac{\pi}{3} + \sin 0 \sin\frac{\pi}{3}$$

$$= 1 \cdot \frac{1}{2} + 0 \cdot \frac{\sqrt{3}}{2} = \frac{1}{2} = \cos\left(-\frac{\pi}{3}\right)$$

This answer is exactly what you get if you use the opposite-angle identity for cosine: $\cos(-\theta) = \cos\theta$ or $\cos\left(-\frac{\pi}{3}\right) = \cos\left(\frac{\pi}{3}\right) = \frac{1}{2}$

In the following example, two negatives make a positive. The angles are positive, but their function values are negative. The two angles in question are α, which is in the fourth quadrant, and β, which is in the third quadrant. The known function values are $\sin\alpha = -\frac{4}{5}$ and $\cos\beta = -\frac{5}{13}$. Find $\cos(\alpha - \beta)$.

1. Find the necessary function values to calculate the difference.

The cosine of the difference of two angles uses both the sine and cosine of each angle involved. You already know the sine of α and cosine of β, so you must determine the cosine of α and the sine of β. Using a Pythagorean identity, you can solve for the missing values:

• First, use the value for $\sin\alpha$ to solve for $\cos\alpha$:

$$\sin^2\alpha + \cos^2\alpha = 1$$

$$\left(-\frac{4}{5}\right)^2 + \cos^2\alpha = 1$$

$$\cos^2\alpha = 1 - \frac{16}{25} = \frac{9}{25}$$

$$\cos\alpha = \pm\sqrt{\frac{9}{25}} = \pm\frac{3}{5}$$

You have to choose between the positive and negative values. Because angle α is in the fourth quadrant, you know that the cosine of α is positive: $\cos\alpha = -\frac{3}{5}$.

• Now use the value for $\cos\beta$ to find $\sin\beta$:

$$\sin^2\beta + \cos^2\beta = 1$$

$$\sin^2\beta + \left(-\frac{5}{13}\right)^2 = 1$$

$$\sin^2\beta = 1 - \frac{25}{169} = \frac{144}{169}$$

$$\sin^2\beta = 1 + \sqrt{\frac{25}{169}} = \pm\frac{12}{13}$$

Again, you choose the correct sign. Angle β is in the third quadrant, where sine is negative, so $\sin\beta = -\frac{12}{13}$.

2. Insert the function values into the identity for the cosine of a difference.

$$\cos(\alpha - \beta) = \cos\alpha \cos\beta + \sin\alpha \sin\beta$$
$$= \left(\frac{3}{5}\right)\left(-\frac{5}{13}\right) + \left(-\frac{4}{5}\right)\left(-\frac{12}{13}\right)$$

3. Simplify the identity and solve for the answer.

$$\cos(\alpha - \beta) = \left(-\frac{15}{65}\right) + \left(\frac{48}{65}\right)$$
$$= \frac{33}{65}$$

In the preceding example, the first angle is in the fourth quadrant, so its measure is between 270 and 360 degrees. The other angle, which is in the third quadrant, is between 180 and 270 degrees. The difference between them could be anywhere between 0 and 180 degrees, meaning that the new angle is either in the first or second quadrant. The answer for the cosine of the difference came out positive. Chapter 9 tells you that the cosine is positive in the first quadrant and negative in the second quadrant, so the difference between the two angles must be somewhere between 0 and 90 degrees, which means that the new angle is in the first quadrant.

Doubling Your Money

Identities for angles that are twice as large as one of the common angles are used a lot in calculus and various math, physics, and science disciplines. These identities allow you to deal with a larger angle in the terms of a smaller and more-manageable one. A *double-angle* function is written, for example, as sin 2θ, cos 2α, or tan $2x$, where 2θ, 2α, and $2x$ are the angle measures and the assumption is that you mean sin (2θ), cos (2α), or tan $(2x)$. In this section, I show you how the double-angle formulas for sine and cosine came to be. I don't go off on a tangent here, but all you need to know is that because tangent is equal to the ratio of sine and cosine, its identity comes from their double-angle identities.

The double-angle identities find the function for twice the angle θ. Note that the cosine function has three different versions of its double-angle identity.

$$\sin 2\theta = 2\sin\theta \cos\theta$$
$$\cos 2\theta = \cos^2\theta - \sin^2\theta$$
$$= 2\cos^2\theta - 1$$
$$= 1 - 2\sin^2\theta$$
$$\tan 2\theta = \frac{2\tan\theta}{1 - \tan^2\theta}$$

One plus one equals two sines

To show you where the double-angle formula for sine comes from, I start with the identity for the sine of a sum, $\sin(\alpha + \beta) = \sin\alpha\cos\beta + \cos\alpha\sin\beta$. If $\alpha = \beta$, then $\alpha + \beta$ becomes $\alpha + \alpha$ or $2(\alpha)$.

I can replace β with α in the formula, giving me

$$\sin(\alpha + \alpha) = \sin\alpha\cos\alpha + \cos\alpha\sin\alpha$$
$$\sin(2\alpha) = 2\sin\alpha\cos\alpha$$

For example, you can use this double-angle identity to find the function value for the sine of 180 degrees.

1. **Determine twice which angle is 180 degrees.**

   Twice 90 is 180, so the choice is 90 degrees.

2. **Substitute the measure into the double-angle identity for sine.**

   $\sin 180° = \sin 2 \cdot 90° = 2 \sin 90° \cos 90°$

3. **Replace the angles with the function values and simplify the answer.**

   $\sin 180° = 2(1)(0) = 0$

But that angle measure is found pretty easily because the terminal side of the angle is on an axis. How about something a bit more challenging. This time, use a double-angle formula to find the sine of 150 degrees.

1. **Determine twice which angle is 150 degrees.**

   Twice 75 is 150, so the choice is 75 degrees. A 75-degree angle isn't one of the basic angles, but you find the value of the sine of 75 degrees earlier in this chapter, in the section "Summing It Up."

2. **Substitute the measure into the double-angle identity for sine.**

   $\sin 150° = \sin 2 \cdot 75° = 2 \sin 75° \cos 75°$

3. **Replace the angles with the function values and simplify the answer.**

   You have $\sin 75° = \dfrac{\sqrt{2} + \sqrt{6}}{4}$ and also need the cosine of the angle.
   A Pythagorean identity comes to the rescue.

$$\sin^2 75° + \cos^2 75° = 1$$

$$\left(\frac{\sqrt{2}+\sqrt{6}}{4}\right)^2 + \cos^2 75° = 1$$

$$\cos^2 75° = 1 - \frac{8+2\sqrt{12}}{16}$$

$$= 1 - \frac{8+4\sqrt{3}}{16}$$

$$= \frac{4}{4} - \frac{2+\sqrt{3}}{4}$$

$$= \frac{2-\sqrt{3}}{4}$$

$$\cos 75° = \sqrt{\frac{2-\sqrt{3}}{4}} = \sqrt{\frac{2-\sqrt{3}}{2}}$$

Whew! Now you see why it's nice to have other options in the form of different types of trig identities.

So, you now have

$$\sin 150° = 2\sin 75° \cos 75°$$

$$= 2\left(\frac{\sqrt{2}+\sqrt{6}}{4}\right)\left(\frac{\sqrt{2-\sqrt{3}}}{2}\right)$$

$$= \frac{\left(\sqrt{2}+\sqrt{6}\right)\sqrt{2-\sqrt{3}}}{4}$$

This can be simplified a bit by multiplying numerator and denominator by the conjugate of the numerator.

$$= \frac{\left(\sqrt{2}+\sqrt{6}\right)\sqrt{2-\sqrt{3}}}{4} \cdot \frac{\sqrt{2}-\sqrt{6}}{\sqrt{2}-\sqrt{6}}$$

$$= \frac{(2-6)\sqrt{2-\sqrt{3}}}{4\left(\sqrt{2}-\sqrt{6}\right)} = \frac{\sqrt{2}-\sqrt{3}}{\sqrt{6}-\sqrt{2}}$$

Still not a pretty sight, but using your calculator to compute the value of that fraction, you get exactly $\frac{1}{2}$. This is *not* the most efficient way to find the sine of 150 degrees, but you see an example of applying a double-angle formula to solve for the value of a trig function.

Three's a crowd

Finding the cosine of twice an angle is easier than finding the other function values, because the cosine offers you three choices. You make your choice depending on what information is available and what looks easiest. To show you where the first of the double-angle identities for cosine comes from, I use the angle-sum identity for cosine. Because the two angles are equal, you can replace β with α, so $\cos(\alpha + \beta) = \cos\alpha\cos\beta - \sin\alpha\sin\beta$ becomes

$$\cos(\alpha + \alpha) = \cos\alpha\cos\alpha - \sin\alpha\sin\alpha$$
$$\cos(2\alpha) = (\cos\alpha)^2 - (\sin\alpha)^2$$
$$= \cos^2\alpha - \sin^2\alpha$$

Unlikely mathematician

Napoleon Bonaparte is best known for his triumphs and trials in French history. But did you know that he was a closet mathematician? He even has a rule named for him. Napoleon is credited with discovering that when you construct equilateral triangles on the sides of any other triangle and then join the centers of those triangles with segments, those segments form another equilateral triangle. Here are some pictures illustrating Napoleon's theorem:

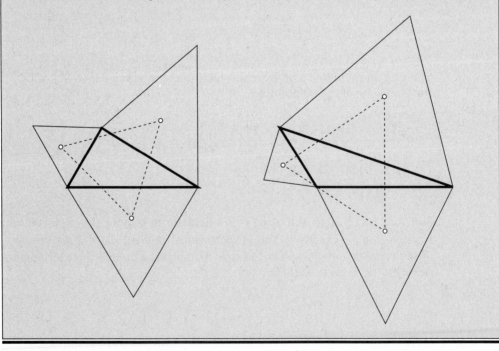

To get the second version, use the first Pythagorean identity, $\sin^2 \alpha + \cos^2 \alpha = 1$. Solving for $\sin^2 \alpha$, you get $\sin^2 \alpha = 1 - \cos^2 \alpha$. Putting this result back into the double-angle identity for cosine and simplifying, you get

$$\cos(2\alpha) = \cos^2 \alpha - \left(1 - \cos^2 \alpha\right)$$
$$= \cos^2 \alpha - 1 + \cos^2 \alpha$$
$$= 2\cos^2 \alpha - 1$$

To find the last version of the double-angle identity for cosine, solve the first Pythagorean identity for $\cos^2 \alpha$, which gives you $\cos^2 \alpha = 1 - \sin^2 \alpha$. Then substitute this result into the first angle-sum identity for cosine:

$$\cos(2\alpha) = \left(1 - \sin^2 \alpha\right) - \sin^2 \alpha$$
$$= 1 - \sin^2 \alpha - \sin^2 \alpha$$
$$= 1 - 2\sin^2 \alpha$$

The biggest advantage to having three different identities for the cosine of a double angle is that you can solve for the cosine with just one other function value. The sum and difference identities for sine and cosine, on the other hand, as well as the double-angle identity for sine, all involve both the sine and cosine of the angles.

Here's an example showing off that advantage. Find $\cos 2\alpha$; the angle α is in the fourth quadrant, and $\sin \alpha = -0.45$.

1. **Choose the appropriate double-angle identity.**

   Because you know the value of the sine, use $\cos 2\alpha = 1 - 2\sin^2 \alpha$.

2. **Insert the given value in the formula and simplify.**

$$\cos 2\alpha = 1 - 2(\sin \alpha)^2 = 1 - 2(-0.45)^2$$
$$= 1 - 2(0.2025) = 1 - 0.4050 = 0.5950$$

The resulting cosine is positive. The cosine is positive in the first and fourth quadrants, so how do you know which of those two quadrants the terminal side of this double angle lies in? Go back to the beginning of the problem — you know that the original angle is in the fourth quadrant. An angle in QIV measures between 270 degrees and 360 degrees. If you double those numbers (because you're working with a double angle), you get 540 degrees and 720 degrees. The angles between those two values lie in the third and fourth quadrants. The cosine is positive in the fourth quadrant, so this double angle lies in the fourth quadrant.

Halving Fun Yet?

The trig identities come in sums, differences, multiples, and halves. With these identities, you can get the value of a sine for a 15-degree angle by using a formula involving half of 30 degrees. You can also get the value of the tangent of a $22\frac{1}{2}$-degree angle by using half of 45 degrees. These identities just create more and more ways to establish an exact value for many of the more commonly used trig functions.

The half-angle identities find the function value for half the measure of angle θ:

$$\sin\frac{\theta}{2} = \pm\sqrt{\frac{1-\cos\theta}{2}}$$

$$\cos\frac{\theta}{2} = \pm\sqrt{\frac{1+\cos\theta}{2}}$$

$$\tan\frac{\theta}{2} = \frac{\sin\theta}{1+\cos\theta} = \frac{1-\cos\theta}{\sin\theta}$$

Notice that the tangent has two versions of its half-angle formula. "Oh, goody!", you say. Just as with the three versions of the cosine's double-angle identities, you get to choose which version is more convenient.

The half-angle identities are a result of taking the double-angle identities and scrunching them around. A more-technical term for *scrunching* is solving for the single angle in a double-angle identity. Here's how the half-angle identity for sine came to be:

1. **Write the double-angle identity for cosine that has just a sine in it.**

   $\cos 2\theta = 1 - 2\sin^2 \theta$

   Using the double-angle identity for cosine works better than the double-angle identity for sine, because the sine formula has both functions on the right side of the equation, and you can't easily get rid of one or the other.

2. **Solve for sin θ. First, get the sin^2 θ term by itself on the left.**

   $\cos 2\theta = 1 - 2\sin^2\theta$

   $2\sin^2\theta = 1 - \cos 2\theta$

3. **Divide each side by 2, and then take the square root of each side.**

   $$\sin^2\theta = \frac{1-\cos 2\theta}{2}$$

   $$\sqrt{\sin^2\theta} = \pm\sqrt{\frac{1-\cos 2\theta}{2}}$$

   $$\sin\theta = \pm\sqrt{\frac{1-\cos 2\theta}{2}}$$

4. Replace 2θ with α and θ with $\frac{\alpha}{2}$.

$$\sin\frac{\alpha}{2} = \pm\sqrt{\frac{1-\cos\alpha}{2}}$$

By switching the letters, you can more easily see the relationship between the two angles, that one is half as big as the other.

You can find the half-angle formula for cosine by using the appropriate choice of cosine's double-angle identity and following similar steps to the preceding ones. With tangent, you use both the sine and cosine identities. But first, what's this business of + or – in the sine and cosine half-angle identities?

Explaining the ±

Trig identities are numerous. Some people say there are too many identities, and others say there just aren't enough. (Like Baby Bear, I think they're just right.) I list the most frequently used identities somewhere in this chapter and in Chapter 11. You may be wondering, however, why these half-angle identities are in a league of their own — some of them have ± in front of them; other identities don't have that lead-in. Continue on, dear reader, for the answer to that nagging question.

What's unique about the half-angle identities for sine and cosine is the fact that the sign attached depends on what quadrant the angle that you're cutting in half is in or how big that angle is. If you want to know the sine of half a 30-degree angle, you look at what quadrant that half is in. Both 30 degrees and its half, 15 degrees, are in the first quadrant, so their sines are both positive. Not so with the sine of 300 degrees and its half, 150 degrees, though. The sine is negative in the fourth quadrant, where the 300-degree angle is located (in standard position), but the sine of 150 degrees is positive, because its terminal side is in the second quadrant. When you apply the half-angle formulas, you have to consider which quadrant each angle is in and apply the appropriate signs.

Half a tangent is double the fun

The half-angle identity for tangent has two versions. Rather than being a nuisance, having more than one option is really rather nice, because you can choose the version that works best for your situation. The half-angle formulas for the tangent involve both sine and cosine, but those functions switch places in the numerator and denominator of the fraction. Sometimes the sine of a function doesn't have a radical in its exact value when the cosine does (or vice versa). Depending on the sine and cosine values, you choose the version of the half-angle tangent identity that'll be easiest to work with after you input the values. The math is easier when you don't have to worry about those radicals in the denominator.

First, where do these half-angle tangent identities come from?

1. **Use the ratio identity for tangent and fill in the half-angle identities for sine and cosine.**

$$\tan\frac{\theta}{2} = \frac{\sin\frac{\theta}{2}}{\cos\frac{\theta}{2}} = \frac{\sqrt{\dfrac{1-\cos\theta}{2}}}{\sqrt{\dfrac{1+\cos\theta}{2}}}$$

   You can leave off the $\pm$ sign because you won't have to choose which sign to use with the tangent identity.

2. **Put the numerator and denominator under the same radical and then simplify the complex fraction.**

$$= \frac{\sqrt{\dfrac{1-\cos\theta}{2}}}{\sqrt{\dfrac{1+\cos\theta}{2}}} = \sqrt{\frac{1-\cos\theta}{2}\cdot\frac{2}{1+\cos\theta}} = \sqrt{\frac{1-\cos\theta}{1+\cos\theta}}$$

3. **Multiply the numerator and denominator by the *conjugate* (same terms, different sign) of the denominator.**

$$= \sqrt{\frac{1-\cos\theta}{1+\cos\theta}\cdot\frac{1-\cos\theta}{1-\cos\theta}} = \sqrt{\frac{(1-\cos\theta)^2}{1+\cos\theta}}$$

4. **Replace the denominator by using the Pythagorean identity and then simplify by putting the radical over the numerator and denominator, individually.**

$$\sqrt{\frac{(1-\cos\theta)^2}{1-\cos^2\theta}} = \sqrt{\frac{(1-\cos\theta)^2}{\sin^2\theta}}$$

$$= \frac{\sqrt{(1-\cos\theta)^2}}{\sqrt{\sin^2\theta}} = \frac{1-\cos\theta}{\sin\theta}$$

To find the other form of this half-angle tangent identity, change Step 3 by multiplying the numerator and denominator of the fraction by the conjugate of the *numerator* instead of the denominator.

Using half-angle identities

By adding, subtracting, or doubling angle measures, you can find lots of exact values of trig functions. This section provides some examples of the types of angles and their functions that you can find with the half-angle identities.

Pi gone wrong

In 1853, William Shanks published his calculation of the decimal value of π to 707 decimal places, which he computed all by hand. Not until 1945 did someone discover that the last 180 digits of this computation were wrong. But an even greater error occurred in 1897. That year, the General Assembly of the State of Indiana enacted Bill Number 246, stating that π was legally equal to 3.2.

Even though you can use a difference identity to find the sine of 15 degrees, you can also use the half-angle identity.

1. **Determine which angle is double the angle you're working with.**

   Half of 30 is 15, so the choice is 30 degrees. Stick to the more-common angles — the ones that have exact values (see Chapter 7) or are multiples of 30 and 45.

2. **Substitute that angle into the half-angle identity for sine.**

$$\sin\frac{\theta}{2} = \pm\sqrt{\frac{1-\cos\theta}{2}}$$
$$\sin 15° = \sin\frac{30}{2} = +\sqrt{\frac{1-\cos 30}{2}}$$

   Because the sine of 15 degrees is a positive value, the sign in front of the radical becomes +.

3. **Fill in the function values and simplify the answer.**

$$\sin 15° = \sqrt{\frac{1-\cos 30}{2}} = \sqrt{\frac{1-\frac{\sqrt{3}}{2}}{2}} = \sqrt{\frac{\frac{2}{2}-\frac{\sqrt{3}}{2}}{2}} = \sqrt{\frac{2-\sqrt{3}}{4}}$$

   The result isn't a particularly pretty value, although beauty is in the eye of the beholder. Some would consider this answer to be wonderful, because it's the exact value and not a decimal approximation.

Now try using the half-angle identity with radians. Find $\tan\frac{\pi}{8}$.

1. **Determine which angle is double the angle you're working with.**

   The angle $\frac{\pi}{4}$ is twice $\frac{\pi}{8}$.

2. **Substitute the angle measure into one of the half-angle tangent identities.**

$$\tan\frac{\theta}{2} = \frac{1-\cos\theta}{\sin\theta}$$

$$\tan\frac{\pi}{8} = \tan\frac{\frac{\pi}{4}}{2} = \frac{1-\cos\frac{\pi}{4}}{\sin\frac{\pi}{4}}$$

3. **Fill in the function values and simplify the answer.**

$$\tan\frac{\pi}{8} = \frac{1-\frac{\sqrt{2}}{2}}{\frac{\sqrt{2}}{2}} = \frac{\frac{2}{2}-\frac{\sqrt{2}}{2}}{\frac{\sqrt{2}}{2}} = \frac{2-\sqrt{2}}{2}\cdot\frac{2}{\sqrt{2}} = \frac{2-\sqrt{2}}{\sqrt{2}}$$

4. **To get the radical out of the denominator, rationalize it by multiplying both parts of the fraction by the conjugate of the denominator.**

$$= \frac{2-\sqrt{2}}{\sqrt{2}}\cdot\frac{\sqrt{2}}{2} = \frac{2\sqrt{2}-2}{2} = \frac{2(\sqrt{2}-1)}{2} = \sqrt{2}-1$$

The other identity for the tangent of a half-angle gives you exactly the same answer. That form isn't any easier, though, because both the sine and cosine of this angle have a radical in them. If the problem involved an angle of 60 degrees, though, the story would be different. The sine of 60 degrees is $\frac{\sqrt{3}}{2}$, and the cosine is $\frac{1}{2}$, which practically begs you to use the form with the cosine in the denominator so you don't have to mess with a radical in the denominator. Both identities work — the one you use is just a matter of personal preference.

Chapter 13

Proving Identities: The Basics

· ·

· ·

*O*ne major aspect that people remember about trigonometry, if they stud-ied it in school, is the time they spent proving identities — making one side of an equation match the other. Some people find proving identities to be the best thing ever — they can't get enough of them. Others, though, find this task of proving identities to be less than exciting — a rite of passage that is best passed by. What you find in this chapter are a game plan and sugges-tions so those who aren't so fond of solving such puzzles so they may begin to actually enjoy the process. For your reading pleasure, I divide this chapter into the methods that work best to prove the different types of identities.

Why do you need to prove identities? Don't you already know that they're correct if they're called identities? Sure you do, but proving them is still helpful down the road when you're solving complex trig problems, because the pro-cess for solving them is all in the preparation. Many of the trig expressions that you use to solve practical problems have rather complicated and nasty-looking terms. By doing substitutions and manipulations with the trig identities, you can make those expressions more usable. All this practice with proving identities prepares you and gives you a heads-up as to what's possible.

Lining Up the Players

Before starting to prove (or solve) identities, you need to look over the different equivalences that you use to solve them.

Here's a quick list for your reference:

Reciprocal Identities

$$\frac{1}{\sin\theta} = \csc\theta \qquad\qquad \frac{1}{\cos\theta} = \sec\theta \qquad\qquad \frac{1}{\tan\theta} = \cot\theta$$

$$\frac{1}{\cot\theta} = \tan\theta \qquad\qquad \frac{1}{\sec\theta} = \cos\theta \qquad\qquad \frac{1}{\csc\theta} = \sin\theta$$

Ratio Identities

$$\tan\theta = \frac{\sin\theta}{\cos\theta}$$

$$\cot\theta = \frac{\cos\theta}{\sin\theta}$$

Opposite-Angle Identities

$$\sin(-\theta) = -\sin\theta$$

$$\cos(-\theta) = \cos\theta$$

$$\tan(-\theta) = -\tan\theta$$

Pythagorean Identities

$$\sin^2\theta + \cos^2\theta = 1 \text{ or } \sin^2\theta = 1 - \cos^2\theta \text{ or } \cos^2\theta = 1 - \sin^2\theta$$

$$\tan^2\theta + 1 = \sec^2\theta \text{ or } \tan^2\theta = \sec^2\theta - 1$$

$$1 + \cot^2\theta = \csc^2\theta \text{ or } \cot^2\theta = \csc^2\theta - 1$$

Sum and Difference Identities

$$\sin(\alpha + \beta) = \sin\alpha\cos\beta + \cos\alpha\sin\beta$$

$$\sin(\alpha - \beta) = \sin\alpha\cos\beta - \cos\alpha\sin\beta$$

$$\cos(\alpha + \beta) = \cos\alpha\cos\beta - \sin\alpha\sin\beta$$

$$\cos(\alpha - \beta) = \cos\alpha\cos\beta + \sin\alpha\sin\beta$$

$$\tan(\alpha + \beta) = \frac{\tan\alpha + \tan\beta}{1 - \tan\alpha\tan\beta}$$

$$\tan(\alpha - \beta) = \frac{\tan\alpha - \tan\beta}{1 + \tan\alpha\tan\beta}$$

Double-Angle Identities

$$\sin 2\theta = 2\sin\theta\cos\theta$$

$$\cos 2\theta = \cos^2\theta - \sin^2\theta = 2\cos^2\theta - 1 = 1 - 2\sin^2\theta$$

$$\tan 2\theta = \frac{2\tan\theta}{1 - \tan^2\theta}$$

Half-Angle Identities

$$\sin\frac{\theta}{2} = \pm\sqrt{\frac{1 - \cos\theta}{2}}$$

$$\cos\frac{\theta}{2} = \pm\sqrt{\frac{1 + \cos\theta}{2}}$$

$$\tan\frac{\theta}{2} = \frac{\sin\theta}{1 + \cos\theta} = \frac{1 - \cos\theta}{\sin\theta}$$

Picking Sides

When you prove identities, you usually work on only one side of the equation or the other — not both at the same time — and for good reason. When you're working in other math areas, such as solving anti-derivatives in calculus, you need to change from one trig expression to another so you can do the problem; such a situation doesn't have any sides, so you need to work on just the one term or expression. By solving trig identities working on just one side or the other is good practice. The good news is that in these problems, you usually get to pick which side.

Here are the guidelines for choosing sides. (Just as in debate, you go for the side you can defend or work with.) You may take your pick based on any one of these options:

- ✔ **Choose the side with the greater number of terms.** Why? Combining terms into one is easier than breaking them apart.

- ✔ **Choose the side with factors that need to be multiplied together.** The reason? Trig functions have a way of merging together because of similar factors in their ratios.

> ✔ **Choose the side with terms other than sine or cosine.** Why? Because all the functions can be written in terms of sine and cosine, so this creates commonalities.
>
> ✔ **Choose the side with fractions that need to added together or subtracted from one another.** The thinking here? Finding common denominators and combining introduces opportunities to apply identities.

Of course, these guidelines aren't all the possibilities, but they give you a good start. Enter the puzzle part of identities. You get to look for clues in each identity to help you decide which side to work on, as in the following examples.

Prove the identity $\cot x \sin x + \tan x \cot x = \cos x + 1$.

1. **Choose the side to work on.**

   The left side has more terms, the first term is a product, and the second term is the product of two reciprocal functions, so this is the best choice.

2. **Use the ratio identity to replace the first cot x and the reciprocal identity to replace the second cot x.**

$$\frac{\cos x}{\sin x} \cdot \sin x + \tan x \cdot \frac{1}{\tan x} = \cos x + 1$$

3. **Multiply the two factors together in the two terms.**

$$\frac{\cos x}{\cancel{\sin x}} \cdot \cancel{\sin x} + \cancel{\tan x} \cdot \frac{1}{\cancel{\tan x}} = \cos x + 1$$
$$\cos x + 1 = \cos x + 1$$

The identity is *proven,* because the two sides are exactly the same. This one went pretty quickly, because the ratio and reciprocal identities were chosen. Another way to approach the problem would be to factor out cot x from the terms on the left and then do some identity-replacing in the terms in the parentheses. This approach isn't as easy, but it can still get the job done.

The next example shows you more techniques.

Prove the identity $\sec x - \sin x \tan x = \cos x$.

1. **Choose the side to work on.**

   The left side has more terms, and two of the functions aren't sine or cosine, so you use two of the guidelines in making this choice.

2. **Use the reciprocal identity to replace sec x and the ratio identity to replace tan x.** This is from the guideline suggesting you change all terms to those involving sine and cosine.

$$\frac{1}{\cos x} - \sin x \cdot \frac{\sin x}{\cos x} = \cos x$$

3. **Multiply the two factors together in the second term. Then combine the two fractions, because they have a common denominator.**

$$\frac{1}{\cos x} - \frac{\sin x}{1} \cdot \frac{\sin x}{\cos x} = \cos x$$

$$\frac{1}{\cos x} - \frac{\sin^2 x}{\cos x} = \cos x$$

$$\frac{1 - \sin^2 x}{\cos x} = \cos x$$

4. **Replace the numerator by using the Pythagorean identity $\sin^2 x + \cos^2 x = 1$, which is also written $\cos^2 x = 1 - \sin^2 x$.**

$$\frac{\cos^2 x}{\cos x} = \cos x$$

5. **Reduce the fraction on the left.**

$$\frac{\cos^{\cancel{2}} x}{\cancel{\cos x}} = \cos x$$

$$\cos x = \cos x$$

The identity is *proven,* because the two sides are exactly the same.

Next, prove the identity

$$\frac{1 - \cos x}{\sin x} + \frac{\sin x}{1 - \cos x} = 2 \csc x.$$

1. **Choose the side to work on.**

   The left side has fractions that you need to add together.

2. **Find the common denominator.**

   The fractions have two different denominators, so multiply each by a fraction that equals 1 — the fraction with the other term's denominator in both the numerator and denominator.

$$\frac{1 - \cos x}{\sin x} \cdot \frac{1 - \cos x}{1 - \cos x} + \frac{\sin x}{1 - \cos x} \cdot \frac{\sin x}{\sin x} = 2 \csc x$$

3. **Simplify the two fractions. Then add them together, because they have the same denominators.**

$$\frac{(1 - \cos x)^2}{\sin x (1 - \cos x)} + \frac{\sin^2 x}{\sin x (1 - \cos x)} = 2 \csc x$$

$$\frac{(1 - \cos x)^2 + \sin^2 x}{\sin x (1 - \cos x)} = 2 \csc x$$

4. **Multiply out the squared binomial in the far-left term of the numerator.**

$$\frac{1-2\cos x+\cos^2 x+\sin^2 x}{\sin x(1-\cos x)}=2\csc x$$

5. **Replace the last two terms in the numerator with 1, using the Pythagorean identity $\sin^2 x + \cos^2 x = 1$.**

$$\frac{1-2\cos x+1}{\sin x(1-\cos x)}=2\csc x$$

6. **Combine the two 1s in the numerator, and then factor out the 2 from each of the terms.**

$$\frac{2-2\cos x}{\sin x(1-\cos x)}=2\csc x$$

$$\frac{2(1-\cos x)}{\sin x(1-\cos x)}=2\csc x$$

7. **Factor out the common multiplier in the numerator and denominator.**

$$\frac{2(1-\cos x)}{\sin x(1-\cos x)}=2\csc x$$

$$\frac{2}{\sin x}=2\csc x$$

8. **Now just use the reciprocal identity, $\frac{1}{\sin x} = \csc x$, to finish up.**

$$2\left(\frac{1}{\sin x}\right)=2\csc x$$

$$2\csc x=2\csc x$$

The next example uses the multiplying-out guideline and the Pythagorean identity to make for a pretty result.

Prove the identity that $\cos x(\sec x - \cos x) = \sin^2 x$.

1. **Decide which side you'll work on.**

The left side just begs to be multiplied out by distributing $\cos x$ over the two terms in the parentheses.

2. **Distribute on the left.**

   $$\cos x \sec x - \cos^2 x = \sin^2 x$$

3. **Cosine and secant are reciprocals, so their product is 1. Replace the term "cos x sec x" with 1.**

   $$\cos x \left(\frac{1}{\cos x} \right) - \cos^2 x = \sin^2 x$$
   $$1 - \cos^2 x = \sin^2 x$$

4. **Now just use the Pythagorean identity to replace the terms on the left.**

   $$\sin^2 x = \sin^2 x$$

Working on Both Sides

As much fun as it is to work on just one side of an identity, sometimes working on both sides at the same time is advantageous and permissible. Working on both sides of an identity is often necessary when you don't have a clearcut way to change one side to match the other. I've even had to resort to working on both sides when it *wasn't* permissible; working backward from one side to the result on the other side can give some valuable clues on how to solve the thing.

With a trig identity, working on both sides isn't really the same as working on both sides of an algebraic equation. In algebra, you can multiply each side by the same number, square both sides, add or subtract the same thing to each side, and so on. When you solve trig identities and equations (see Chapter 17), you can use all those algebra rules *plus* you can do substitutions with the various trig identities when you need them. You can even insert a different identity on each side — the one big advantage of working on both sides of a trig identity.

This first example is rather basic, but it gets the idea across. Solve the identity $\dfrac{\sin\theta}{\csc\theta} + \dfrac{\cos\theta}{\sec\theta} = \tan\theta\cot\theta$ by working on both sides.

1. **Replace the two denominators of the fractions with their reciprocal identities. Also replace the cotangent on the right with its reciprocal.**

   $$\frac{\sin\theta}{\frac{1}{\sin\theta}} + \frac{\cos\theta}{\frac{1}{\cos\theta}} = \tan\theta \left(\frac{1}{\tan\theta} \right)$$

2. **Simplify the two fractions on the left by flipping the denominators and multiplying them by their numerators. Then multiply the two factors on the right together.**

$$\sin\theta \cdot \frac{\sin\theta}{1} + \cos\theta \cdot \frac{\cos\theta}{1} = 1$$

$$\frac{\sin\theta}{1} \cdot \frac{\sin\theta}{1} + \frac{\cos\theta}{1} \cdot \frac{\cos\theta}{1} = 1$$

$$\sin^2\theta + \cos^2\theta = 1$$

3. **Replace the sum on the left using the Pythagorean identity.**

You end up with $1 = 1$.

In the next example, you change everything to sines and cosines. Prove the identity $\frac{\csc x}{\cos x} = \cot x + \tan x$.

1. **Change the functions to their equivalences by using the reciprocal and ratio identities.**

$$\frac{\frac{1}{\sin x}}{\cos x} = \frac{\cos x}{\sin x} + \frac{\sin x}{\cos x}$$

2. **On the left, write the denominator as a fraction and then flip it and multiply it by the numerator. On the right, multiply each fraction by a fraction equal to 1 (by using the other fraction's denominator) to get common denominators for all the fractions.**

$$\frac{\frac{1}{\sin x}}{\frac{\cos x}{1}} = \frac{\cos x}{\sin x} + \frac{\sin x}{\cos x}$$

$$\frac{1}{\sin x} \cdot \frac{1}{\cos x} = \frac{\cos x}{\sin x} \cdot \frac{\cos x}{\cos x} + \frac{\sin x}{\cos x} \cdot \frac{\sin x}{\sin x}$$

3. **Simplify the multiplied fractions. Add the two fractions on the right together.**

$$\frac{1}{\sin x \cos x} = \frac{\cos^2 x}{\sin x \cos x} + \frac{\sin^2 x}{\sin x \cos x}$$

$$= \frac{\cos^2 x + \sin^2 x}{\sin x \cos x}$$

4. **Replace the numerator on the right with the value from the Pythagorean identity.**

$$\frac{1}{\sin x \cos x} = \frac{1}{\sin x \cos x}$$

The music of the spheres

Pythagoras is best known for his theorem, which defines the relationships among the lengths of a right triangle's sides, but his second most well-known contribution to humanity is his discovery of the mathematical basis of the musical scale. He found that a connection exists between musical harmony — the stuff that sounds good — and whole numbers. If you pluck a taut string, listen to the note, and then pluck a string twice as long and equally taut, you hear a note one octave below the first note. You can also go down the scale by increasing the length of the taut string in smaller increments.

Pythagoras believed that whole-number relationships represent all harmony, all beauty, and all nature. He extended this theory to the orbits of the planets and believed that as the planets move through space, they must give off a heavenly whole-number harmony. Hence the term the *music of the spheres* (in one of my favorite songs from *Les Misérables*).

When working on both sides, you're done when the two sides read the same. It's different from working on one side, where you keep something from the original equation.

This last example requires a little creativity to get the job done. But working on both sides still works best when solving $\frac{1+\cot\alpha}{\cot\alpha} = \tan\alpha + \csc^2\alpha - \cot^2\alpha$.

1. **Split up the fraction on the left by writing each term in the numerator over the denominator.**

$$\frac{1}{\cos\alpha} + \frac{\cot\alpha}{\cot\alpha} = \tan\alpha + \csc^2\alpha - \cot^2\alpha$$

2. **Reduce the second fraction to 1.**

$$\frac{1}{\cot\alpha} + 1 = \tan\alpha + \csc^2\alpha - \cot^2\alpha$$

3. **Now replace the $\csc^2\alpha$ on the right with its equivalent by using the Pythagorean identity.**

$$\frac{1}{\cot\alpha} + 1 = \tan\alpha + \left(1 + \cot^2\alpha\right) - \cot^2\alpha$$

4. **Simplify the terms on the right after dropping the parentheses — two of the terms are opposites of one another.**

$$\frac{1}{\cot\alpha} + 1 = \tan\alpha + 1 + \cot^2\alpha - \cot^2\alpha$$
$$= \tan\alpha + 1$$

5. **Replace the fraction on the left by using the reciprocal identity.**

$$\tan \alpha + 1 = \tan \alpha + 1$$

Going Back to Square One

With some identities, which side you should work on or what you should do with either or both sides isn't clear. And in some instances, you're faced with such a conglomeration of functions that figuring out what's going on is darn near impossible. Other times, the different terms have different powers of the same function. In such cases, simplifying matters either by changing everything to sines and cosines or by factoring out some function may be your best bet.

Changing to sines and cosines

In this first example, you can use either reciprocal or ratio identities, depending on which side you're going to work on, to change everything to sines and cosines. Here's how I'd solve the identity $\tan \theta + \cot \theta = \csc \theta \sec \theta$:

1. **Going with the guideline to work on the side with the greatest number of terms, replace the two terms on the left by using ratio identities.**

$$\frac{\sin\theta}{\cos\theta} + \frac{\cos\theta}{\sin\theta} = \csc\theta \sec\theta$$

2. **To get a common denominator, multiply both terms on the left by fractions equal to 1 (by using the other term's denominator).**

$$\frac{\sin\theta}{\cos\theta} \cdot \frac{\sin\theta}{\sin\theta} + \frac{\cos\theta}{\sin\theta} \cdot \frac{\cos\theta}{\cos\theta} = \csc\theta \sec\theta$$

3. **Simplify the fractions and then add them together, because now they have a common denominator.**

$$\frac{\sin^2\theta}{\cos\theta \sin\theta} + \frac{\cos^2\theta}{\sin\theta \cos\theta} = \csc\theta \sec\theta$$

$$\frac{\sin^2\theta + \cos^2\theta}{\cos\theta \sin\theta} = \csc\theta \sec\theta$$

4. **Replace the numerator on the left with its equivalent by using the Pythagorean identity.**

$$\frac{1}{\cos\theta\sin\theta} = \csc\theta\sec\theta$$

5. **Now use the reciprocal identities on the terms in the denominator and then flip each fraction and multiply.**

$$\frac{1}{\frac{1}{\sec\theta}\cdot\frac{1}{\csc\theta}} = \csc\theta\sec\theta$$

$$\frac{\sec\theta}{1}\cdot\frac{\csc\theta}{1} = \csc\theta\sec\theta$$

$$\sec\theta\csc\theta = \sec\theta\csc\theta$$

In the next example, only two terms aren't already written as sines, so replacing those two with terms in sines just seems natural when solving

$$\frac{\sin x + 8\csc x}{\sin x + 4\csc x} = \frac{\sin^2 x + 8}{\sin^2 x + 4}.$$

1. **Change the two cosecants on the left by using the reciprocal identity.**

$$\frac{\sin x + 8\left(\dfrac{1}{\sin x}\right)}{\sin x + 4\left(\dfrac{1}{\sin x}\right)} = \frac{\sin^2 x + 8}{\sin^2 x + 4}$$

$$\frac{\sin x + \dfrac{8}{\sin x}}{\sin x + \dfrac{4}{\sin x}} = \frac{\sin^2 x + 8}{\sin^2 x + 4}$$

2. **Multiply each term in the numerator and denominator of the left-hand side by sin x. This action amounts to multiplying by sine over sine, or by 1.**

$$\frac{\sin x}{\sin x}\cdot\frac{\sin x + \dfrac{8}{\sin x}}{\sin x + \dfrac{4}{\sin x}} = \frac{\sin^2 x + 8}{\sin^2 x + 4}$$

$$\frac{\sin x\left(\sin x + \dfrac{8}{\sin x}\right)}{\sin x\left(\sin x + \dfrac{4}{\sin x}\right)} = \frac{\sin^2 x + 8}{\sin^2 x + 4}$$

$$\frac{\sin x\cdot\sin x + \sin x\cdot\dfrac{8}{\sin x}}{\sin x\cdot\sin x + \sin x\cdot\dfrac{4}{\sin x}} = \frac{\sin^2 x + 8}{\sin^2 x + 4}$$

3. Simplify the numerator and denominator.

$$\frac{\sin^2 x + \sin x \cdot \dfrac{8}{\sin x}}{\sin^2 x + \sin x \cdot + \dfrac{4}{\sin x}} = \frac{\sin^2 x + 8}{\sin^2 x + 4}$$

$$\frac{\sin^2 x + 8}{\sin^2 x + 4} = \frac{\sin^2 x + 8}{\sin^2 x + 4}$$

This last example has so many different functions and terms that figuring out where to start almost seems impossible. Although you have other ways to approach it, I change the fraction on the left to all sines and cosines. If you want to see another way to solve an identity like this one, refer to the "Finding a common denominator" section, in Chapter 14.

Solve the identity

$$\frac{1 + \sec x}{\tan x} - \frac{\tan x}{\sec x} = \cot x (1 + \cos x).$$

1. On the left side, change the secants by using the reciprocal identity and the tangents by using the ratio identity.

$$\frac{1 + \dfrac{1}{\cos x}}{\dfrac{\sin x}{\cos x}} - \frac{\dfrac{\sin x}{\cos x}}{\dfrac{1}{\cos x}} = \cot x (1 + \cos x)$$

2. On the left, multiply each term in the numerator and denominator by cos x and simplify all the terms.

$$\frac{\cos x}{\cos x} \cdot \frac{1 + \dfrac{1}{\cos x}}{\dfrac{\sin x}{\cos x}} - \frac{\cos x}{\cos x} \cdot \frac{\dfrac{\sin x}{\cos x}}{\dfrac{1}{\cos x}} = \cot x (1 + \cos x)$$

$$\frac{\cos x \left(1 + \dfrac{1}{\cos x}\right)}{\cos x \left(\dfrac{\sin x}{\cos x}\right)} - \frac{\cos x \left(\dfrac{\sin x}{\cos x}\right)}{\cos x \left(\dfrac{1}{\cos x}\right)} = \cot x (1 + \cos x)$$

$$\frac{\cos x \cdot 1 + \cos x \cdot \dfrac{1}{\cos x}}{\cos x \cdot \dfrac{\sin x}{\cos x}} - \frac{\cos x \cdot \dfrac{\sin x}{\cos x}}{\cos x \cdot \dfrac{1}{\cos x}} = \cot x (1 + \cos x)$$

$$\frac{\cos x + \cos x \cdot \dfrac{1}{\cos x}}{\cos x \cdot \dfrac{\sin x}{\cos x}} - \frac{\cos x \cdot \dfrac{\sin x}{\cos x}}{\cos x \cdot \dfrac{1}{\cos x}} = \cot x (1 + \cos x)$$

From that mess, you get $\dfrac{\cos x + 1}{\sin x} - \dfrac{\sin x}{1} = \cot x (1 + \cos x)$.

3. **Find a common denominator for the two fractions on the left, add the fractions together, and simplify the result.**

$$\frac{\cos x + 1}{\sin x} - \frac{\sin x}{\sin x} \cdot \frac{\sin x}{1} = \cot x (1 + \cos x)$$

$$\frac{\cos x + 1}{\sin x} - \frac{\sin^2 x}{\sin x} = \cot x (1 + \cos x)$$

$$\frac{\cos x + 1 - \sin^2 x}{\sin x} = \cot x (1 + \cos x)$$

4. **Now replace the $\sin^2 x$ in the numerator with its equivalent by using the Pythagorean identity, and simplify.**

$$\frac{\cos x + 1 - \left(1 - \cos^2 x\right)}{\sin x} = \cot x (1 + \cos x)$$

$$\frac{\cos x + 1 - 1 + \cos^2 x}{\sin x} = \cot x (1 + \cos x)$$

$$\frac{\cos x + \cos^2 x}{\sin x} = \cot x (1 + \cos x)$$

5. **Factor a $\cos x$ from each term in the numerator.**

$$\frac{\cos x (1 + \cos x)}{\sin x} = \cot x (1 + \cos x)$$

6. **Finally, split the two factors in the numerator into two fractions that are multiplied by each other. Then replace $\dfrac{\cos x}{\sin x}$ by using the ratio identity.**

$$\frac{\cos x}{\sin x} \cdot \frac{(1 + \cos x)}{1} = \cot x (1 + \cos x)$$

$$\cot x (1 + \cos x) = \cot x (1 + \cos x)$$

Factoring

The clue you'll get that says you should factor an identity is when powers of a particular function or repeats of that same function are in all the terms on one side of the identity.

For example, the identity $\sin^4 \theta + 2 \sin^2 \theta \cos^2 \theta + \cos^4 \theta = 1$ has three terms on the left that you can factor, because they're the result of squaring a binomial. The pattern you need is the algebraic equation for the square of a binomial: $a^2 + 2ab + b^2 = (a + b)^2$.

1. **Factor the expression on the left as the square of a binomial.**

   $(\sin^2 \theta + \cos^2 \theta)^2 = 1$

2. **Now just replace the expression in the parentheses with its equivalent by using the Pythagorean identity.**

   $(1)^2 = 1$

The preceding example was really simple — as long as you recognized the pattern of the square of a binomial. It'd be another thing altogether if you went off on some tangent (pardon the pun).

In the next example, the factoring occurs in the numerator of the fraction, where powers of $\sin x$ appear. Solve the identity $\dfrac{\sin x - \sin^3 x}{\cos x} = \tan x \cos^2 x$.

1. **Factor $\sin x$ out of each term in the numerator.**

   $\dfrac{\sin x \left(1 - \sin^2 x\right)}{\cos x} = \tan x \cos^2 x$

2. **Replace the expression in the parentheses with its equivalent by using the Pythagorean identity.**

   $\dfrac{\sin x \left(\cos^2 x\right)}{\cos x} = \tan x \cos^2 x$

3. **Now split up the fraction into a product of two fractions, carefully arranging the factors in the numerator and denominator.**

   $\dfrac{\sin x}{\cos x} \cdot \dfrac{\left(\cos^2 x\right)}{1} = \tan x \cos^2 x$

4. **Replace the first fraction with $\tan x$ by using the ratio identity.**

   $\tan x \cos^2 x = \tan x \cos^2 x$

This last example requires factoring by using the difference between two squares. The pattern here is the algebraic equation $a^2 - b^2 = (a - b)(a + b)$ or $a^4 - b^4 = (a^2 - b^2)(a^2 + b^2)$. Solve the identity $\csc^2 \theta + \cot^2 \theta = \csc^4 \theta - \cot^4 \theta$.

1. **Factor the two terms on the right by using the difference-of-squares pattern.**

$$\csc^2 \theta + \cot^2 \theta = (\csc^2 \theta - \cot^2 \theta)(\csc^2 \theta + \cot^2 \theta)$$

2. **In the left set of parentheses only, replace $\csc^2 \theta$ with its equivalent in the Pythagorean identity.**

You want to keep the two terms in the right parentheses as written.

$$\csc^2 \theta + \cot^2 \theta = (1 + \cot^2 \theta - \cot^2 \theta)(\csc^2 \theta + \cot^2 \theta)$$

3. **Now simplify the expression, getting rid of the two opposites.**

$$\csc^2 \theta + \cot^2 \theta = \left(1 + \cancel{\cot^2 \theta} - \cancel{\cot^2 \theta}\right)\left(\csc^2 \theta + \cot^2 \theta\right)$$
$$= (1)\left(\csc^2 \theta + \cot^2 \theta\right)$$

Using a little bit of both

Just when you thought that proving identities couldn't be much more fun than what you've seen, you now find that the examples in this section involve both changing the terms to sines and cosines as well as factoring. The hardest part is deciding what to do first.

In this first example, your work goes more smoothly if you change everything to sines and cosines first. (Plus you may not recognize right away that the expression on the left is the result of squaring a binomial.) Solve the identity

$$\csc^2 \theta - 2\csc \theta \cot \theta + \cot^2 \theta = \frac{1 - \cos \theta}{1 + \cos \theta}.$$

1. **Change the terms on the left to sines and cosines by using reciprocal and ratio identities, and then simplify the fractions.**

$$\frac{1}{\sin^2 \theta} - \frac{2}{\sin \theta} \cdot \frac{\cos \theta}{\sin \theta} + \frac{\cos^2 \theta}{\sin^2 \theta} = \frac{1 - \cos \theta}{1 + \cos \theta}$$

$$\frac{1}{\sin^2 \theta} - \frac{2\cos \theta}{\sin^2 \theta} + \frac{\cos^2 \theta}{\sin^2 \theta} = \frac{1 - \cos \theta}{1 + \cos \theta}$$

2. **Add the three fractions on the left together, because they have the same denominator.**

$$\frac{1 - 2\cos \theta + \cos^2 \theta}{\sin^2 \theta} = \frac{1 - \cos \theta}{1 + \cos \theta}$$

3. **Replace the denominator of the fraction by using the Pythagorean identity.**

   You usually don't go from a simple one term to two terms, but, looking ahead, you see that you need $1 + \cos\theta$ in the denominator, so this seems like a good idea.

   $$\frac{1-2\cos\theta+\cos^2\theta}{1-\cos^2\theta} = \frac{1-\cos\theta}{1+\cos\theta}$$

4. **Factor the numerator as the square of a binomial; factor the denominator as the difference of two squares.**

   $$\frac{(1-\cos\theta)^2}{(1-\cos\theta)(1+\cos\theta)} = \frac{1-\cos\theta}{1+\cos\theta}$$

5. **Factor out the common binomial in the numerator and denominator.**

   $$\frac{(1-\cos\theta)^{\cancel{2}}}{\cancel{(1-\cos\theta)}(1+\cos\theta)} = \frac{1-\cos\theta}{1+\cos\theta}$$

   $$\frac{1-\cos\theta}{1+\cos\theta} = \frac{1-\cos\theta}{1+\cos\theta}$$

   You find a lot of algebra in trigonometry!

In the next example, you see how to first do the factoring and then go to the basics. Solve the identity $\dfrac{\sin\theta}{\cot\theta - \cot\theta\cos^2\theta} - \sec\theta = 0$.

1. **Factor cot θ out of each term in the denominator of the fraction.**

   $$\frac{\sin\theta}{\cot\theta(1-\cos^2\theta)} - \sec\theta = 0$$

2. **Replace the value in the parentheses with its equivalent by using the Pythagorean identity.**

   $$\frac{\sin\theta}{\cot\theta(\sin^2\theta)} - \sec\theta = 0$$

3. **Write everything in terms of sine and cosine by using reciprocal and ratio identities.**

   $$\frac{\sin\theta}{\frac{\cos\theta}{\sin\theta}(\sin^2\theta)} - \frac{1}{\cos\theta} = 0$$

4. Reduce the fractions in the denominator and simplify.

$$\frac{\sin\theta}{\dfrac{\cos\theta}{\cancel{\sin\theta}}\left(\dfrac{\sin^2\theta}{1}\right)} - \frac{1}{\cos\theta} = 0$$

$$\frac{\sin\theta}{\cos\theta\sin\theta} - \frac{1}{\cos\theta} = 0$$

5. Divide out sin θ from the first fraction and simplify.

$$\frac{\cancel{\sin\theta}}{\cos\theta\,\cancel{\sin\theta}} - \frac{1}{\cos\theta} = 0$$

$$\frac{1}{\cos\theta} - \frac{1}{\cos\theta} = 0$$

$$0 = 0$$

Chapter 14

Sleuthing Out Identity Solutions

In This Chapter

▶ Handling fractions with care

▶ Maneuvering with handy algebraic tricks

▶ Getting creative with math operations to prove identities

*P*roving a trig identity can be a simple chore, or it can be a challenge. The nice thing about an identity is that you *know* that it can be proved — it's an identity, for goodness sake. Some identities seem to just call out with the methods needed to prove them. "Look at me! Look at the three terms on the right that begging to be combined!" Other identities just sit there — daring you to do anything about them.

In this chapter, you find more techniques and suggestions for handling identities. You always want to find the simplest way, first . . . if there is a simplest way. If the easy road fails you, then get on this super highway of trigonometric maneuvers.

Fracturing Fractions

The ratio and reciprocal identities involve fractions. The half-angle identities use fractions. You just can't get away from them. Actually, an identity with fractions can work to your advantage. You can work toward getting rid of the fraction and, in the process, solve the problem. Some of the main techniques for working with fractions in identities are either to break them up into separate terms or to go in the other direction and find a common denominator. You'll find some examples of using a common denominator in Chapter 13 — and even more in this chapter.

Breaking up is hard to do

This section's heading is very misleading. Breaking up fractions really isn't all that hard to do. In fact, when you can do it, breaking up fractions is one of the most productive ways to solve identities. The trick is to break them up correctly. You can break up a fraction with several terms in the numerator and one in the denominator — but not the other way around.

Correct: $\dfrac{1+a-b}{7} = \dfrac{1}{7} + \dfrac{a}{7} - \dfrac{b}{7}$

Incorrect: $\dfrac{3+a-z}{2+b+c} \neq \dfrac{3}{2} + \dfrac{a}{b} - \dfrac{z}{c}$

Now apply this breaking up of fractions to a trig identity. In this first example, the fraction on the left has just one term in the denominator. Solve the identity $\dfrac{\sin x + \cot x}{\cos x} = \tan x + \csc x$. You have the hint that breaking up fractions will work, because you see two terms on the right. You want to have two terms on the left, also.

1. **Break up the fraction by writing each term in the numerator on the left over the denominator.**

   $\dfrac{\sin x}{\cos x} + \dfrac{\cot x}{\cos x} = \tan x + \csc x$

2. **Rewrite cot x by using the ratio identity.**

   $\dfrac{\sin x}{\cos x} + \dfrac{\frac{\cos x}{\sin x}}{\cos x} = \tan x + \csc x$

3. **Simplify the complex fraction by flipping the denominator and multiplying it times the numerator. Then reduce the result.**

   $\dfrac{\sin x}{\cos x} + \dfrac{\cancel{\cos x}}{\sin x} \cdot \dfrac{1}{\cancel{\cos x}} = \tan x + \csc x$

   $\dfrac{\sin x}{\cos x} + \dfrac{1}{\sin x} = \tan x + \csc x$

4. **Replace the first fraction by using the ratio identity for tangent and the second fraction by using the reciprocal identity for cosecant.**

   $\tan x + \csc x = \tan x + \csc x$

The next example doesn't give you the hint about matching the number of terms on each side. Both sides have the same number of terms already. What catches your eye is the factoring possibilities if the left side is written as two fractions. You can break up fractions that have more than one factor (but only one term) in the denominator by carrying them both along. For example, solve the identity $\dfrac{\cot x - \cos x}{\cot x \cos x} = \dfrac{1 - \sin x}{\cos x}$

1. **Break up the fraction on the left by writing each term in the numerator over the entire denominator.**

   $\dfrac{\cot x}{\cot x \cos x} - \dfrac{\cos x}{\cot x \cos x} = \dfrac{1 - \sin x}{\cos x}$

2. Reduce each fraction on the left side.

$$\frac{\cancel{\cot x}}{\cancel{\cot x}\cos x} - \frac{\cancel{\cos x}}{\cot x\,\cancel{\cos x}} = \frac{1-\sin x}{\cos x}$$

$$\frac{1}{\cos x} - \frac{1}{\cot x} = \frac{1-\sin x}{\cos x}$$

3. Rewrite cot x in the second denominator by using the ratio identity. Then simplify the complex fraction by flipping the denominator and multiplying.

$$\frac{1}{\cos x} - \frac{1}{\dfrac{\cos x}{\sin x}} = \frac{1-\sin x}{\cos x}$$

$$\frac{1}{\cos x} - 1\left(\frac{\sin x}{\cos x}\right) = \frac{1-\sin x}{\cos x}$$

4. The two fractions on the left now have the same denominator. Rewrite the left side as all one fraction. (What was fractured will now be rejoined.)

$$\frac{1-\sin x}{\cos x} = \frac{1-\sin x}{\cos x}$$

The next example shows you a proof by breaking a fraction after involving Pythagoras. Prove the identity $\dfrac{1+\cos x(\sin x - \cos x)}{\sin x} = \sin x + \cos x$.

1. First distribute cos x in the numerator over the two terms in the binomial.

$$\frac{1+\cos x \sin x - \cos^2 x}{\sin x} = \sin x + \cos x$$

2. Regroup the terms so that $1 - \cos^2 x$ appear together, suggesting a substitution using a Pythagorean triple.

$$\frac{1-\cos^2 x + \cos x \sin x}{\sin x} = \sin x + \cos x$$

$$\frac{\sin^2 x + \cos x \sin x}{\sin x} = \sin x + \cos x$$

3. Break up the fraction on the left, writing each term over the denominator.

$$\frac{\sin^2 x}{\sin x} + \frac{\cos x \sin x}{\sin x} = \sin x + \cos x$$

4. Reduce the fractions.

$$\frac{\sin^2 x}{\cancel{\sin x}} + \frac{\cos x \cancel{\sin x}}{\cancel{\sin x}} = \sin x + \cos x$$

$$\sin x + \cos x = \sin x + \cos x$$

Finding a common denominator

Fractions are your friends. You may find this unbelievable, but the more you work with trig functions, the more you're swayed to my way of thinking. Finding a common denominator to combine fractions often paves the way to solving an identity.

In the identity $\dfrac{1-\sin\theta}{\cos\theta} - \dfrac{\cos\theta}{1+\sin\theta} = 0$, the two denominators on the left have nothing in common, so you multiply each fraction by the other's denominator — or, rather, by that denominator over itself, which equals 1.

1. **Multiply each fraction on the left by an equivalent of 1 to create a common denominator.**

$$\frac{1-\sin\theta}{\cos\theta}\cdot\frac{1+\sin\theta}{1+\sin\theta} - \frac{\cos\theta}{1+\sin\theta}\cdot\frac{\cos\theta}{\cos\theta} = 0$$

2. **Multiply the fractions together and simplify the numerators. Leave the denominator alone.**

$$\frac{(1-\sin\theta)(1+\sin\theta)}{\cos\theta(1+\sin\theta)} - \frac{\cos\theta\cos\theta}{\cos\theta(1+\sin\theta)} = 0$$

$$\frac{1-\sin^2\theta}{\cos\theta(1+\sin\theta)} - \frac{\cos^2\theta}{\cos\theta(1+\sin\theta)} = 0$$

3. **Replace the first numerator with its equivalent by using the Pythagorean identity.**

The fractions are opposites of one another.

$$\frac{\cos^2\theta}{\cos\theta(1+\sin\theta)} - \frac{\cos^2\theta}{\cos\theta(1+\sin\theta)} = 0$$

$$0 = 0$$

In the "Changing to sines and cosines" section in Chapter 13, I did a problem using that method and mentioned that you had another option — finding a common denominator. You have to decide which way you think is better. You may even be able to find an easier way to do this proof. Here you go: Prove the identity $\dfrac{1+\sec x}{\tan x} - \dfrac{\tan x}{\sec x} = \cot x(1+\cos x)$ by finding a common denominator.

You can see why the option of changing to all sines and cosines may have been your first choice.

1. **Multiply each fraction on the left by the equivalent of 1, creating a common denominator.**

$$\frac{1+\sec x}{\tan x}\cdot\frac{\sec x}{\sec x} - \frac{\tan x}{\sec x}\cdot\frac{\tan x}{\tan x} = \cot x(1+\cos x)$$

2. Simplify the numerators by multiplying out the fractions.

$$\frac{\sec x + \sec^2 x}{\tan x \sec x} - \frac{\tan^2 x}{\tan x \sec x} = \cot x (1 + \cos x)$$

3. Replace tan² *x* in the second fraction with its equivalent by using the Pythagorean identity; then combine the two numerators.

$$\frac{\sec x + \sec^2 x}{\tan x \sec x} - \frac{\sec^2 x - 1}{\tan x \sec x} = \cot x (1 + \cos x)$$

$$\frac{\sec x + \sec^2 x - \sec^2 x + 1}{\tan x \sec x} = \cot x (1 + \cos x)$$

4. Simplify the numerator; then rewrite the left side as the product of two fractions.

$$\frac{\sec x + \cancel{\sec^2 x} - \cancel{\sec^2 x} + 1}{\tan x \sec x} = \cot x (1 + \cos x)$$

$$\frac{\sec x + 1}{\tan x \sec x} = \cot x (1 + \cos x)$$

$$\frac{1}{\tan x} \cdot \frac{\sec x + 1}{\sec x} = \cot x (1 + \cos x)$$

Note that this rewriting as the *product* of two fractions isn't breaking up the fraction — you still have just one term.

5. Multiply the by cos *x* divided by cos *x*, which is equivalent to 1.

You only need to multiply one factor in the numerator and one in the denominator by cos *x*. You cleverly choose the factors with sec *x* in them.

$$\frac{1}{\tan x} \cdot \frac{\sec x + 1}{\sec x} \cdot \frac{\cos x}{\cos x} = \cot x (1 + \cos x)$$

$$\frac{1}{\tan x} \left(\frac{\sec x + 1}{\sec x} \cdot \frac{\cos x}{\cos x} \right) = \cot x (1 + \cos x)$$

6. Multiply out the second fraction, distributing through the numerator.

$$\frac{1}{\tan x} \left(\frac{\sec x \cos x + 1 \cdot \cos x}{\sec x \cos x} \right) = \cot x (1 + \cos x)$$

7. Because cos *x* and sec *x* are reciprocals, their product is 1; substitute 1 for sec *x* cos *x* in both the numerator and the denominator.

$$\frac{1}{\tan x} \left(\frac{1 + \cos x}{1} \right) = \cot x (1 + \cos x)$$

8. Replace the reciprocal of tan *x* with cot *x*.

$$\cot x (1 + \cos x) = \cot x (1 + \cos x)$$

Using Tricks of the Trig Trade

When proving identities, sometimes the best way to handle them just leaps out at you — and sometimes the best way just stays in hiding. Usually, you can solve an identity in more than one way — the best way, the almost-as-good way, the reasonable way, and the absolutely dreadful way. The best way is the quickest and most efficient. But sometimes you have to pull something out of your hat to accomplish the task of solving a particular identity. You've already seen one little trick: multiplying a term by 1. Well, you multiply by sine over sine or some such arrangement, but it's still just multiplying by 1. Some additional little tricks amount to nothing more than multiplying a fraction by 1 in the form of a conjugate or squaring both sides of the identity.

Multiplying by a conjugate

First, what in the world is a *conjugate?* In mathematics, a conjugate consists of the same two terms as the first expression, separated by the opposite sign. For instance, the conjugate of $x + \sqrt{y}$ is $x - \sqrt{y}$. In trig, especially, multiplying the numerator and denominator of a fraction by a conjugate can create some really nice results.

Multiplying by a conjugate is a quick, easy way of solving the identity

$$\frac{1}{\sec x - \tan x} = \tan x + \sec x.$$

1. **Multiply the numerator and denominator of the fraction on the left by the conjugate of the denominator.**

$$\frac{1}{\sec x - \tan x} \cdot \frac{\sec x + \tan x}{\sec x + \tan x} = \tan x + \sec x$$

2. **The two denominators multiplied together are the difference of two squares.**

$$\frac{\tan x + \sec x}{\sec^2 x - \tan^2 x} = \tan x + \sec x$$

3. **Replace $\sec^2 x$ in the denominator with its equivalent by using the Pythagorean identity.**

$$\frac{\tan x + \sec x}{\tan^2 x + 1 - \tan^2 x} = \tan x + \sec x$$

4. **Simplify the denominator by canceling out the two opposites.**

$$\frac{\tan x + \sec x}{\cancel{\tan^2 x} + 1 - \cancel{\tan^2 x}} = \tan x + \sec x$$

$$\tan x + \sec x = \tan x + \sec x$$

In the next example, you have to decide which fraction to multiply the conjugate by. I choose the fraction on the right, because I see the conjugate of the numerator on the right in the denominator on the left. Solve the identity $\dfrac{\tan x}{1 + \cos x} = \dfrac{1 - \cos x}{\sin x \cos x}$.

1. **Multiply the numerator and denominator of the fraction on the right by the conjugate of the numerator.**

$$\frac{\tan x}{1 + \cos x} = \frac{1 - \cos x}{\sin x \cos x} \cdot \frac{1 + \cos x}{1 + \cos x}$$

2. **Multiply the fractions together, keeping the parentheses in the denominator.**

$$\frac{\tan x}{1 + \cos x} = \frac{1 - \cos^2 x}{\sin x \cos x (1 + \cos x)}$$

3. **Substitute the equivalent from the Pythagorean identity in the numerator of the fraction on the right. Then reduce that fraction.**

$$\frac{\tan x}{1 + \cos x} = \frac{\sin^2 x}{\sin x \cos x (1 + \cos x)}$$

$$\frac{\tan x}{1 + \cos x} = \frac{\sin^2 x}{\cancel{\sin x} \cos x (1 + \cos x)}$$

$$\frac{\tan x}{1 + \cos x} = \frac{\sin x}{\cos x (1 + \cos x)}$$

4. **Rewrite the fraction on the right as a product of two fractions, carefully arranging the factors.**

$$\frac{\tan x}{1 + \cos x} = \frac{\sin x}{\cos x} \cdot \frac{1}{(1 + \cos x)}$$

5. **Replace the first fraction on the right with its ratio-identity equivalent. Rewrite the expression as one fraction.**

$$\frac{\tan x}{1+\cos x} = \tan x \cdot \frac{1}{(1+\cos x)}$$

$$\frac{\tan x}{1+\cos x} = \frac{\tan x}{1+\cos x}$$

The half-angle identity for the tangent function has two different forms. Multiplying by the conjugate is a good method for showing that these two forms are equivalent. In this example, I prove that the two half-angle identities are equivalent, $\dfrac{\sin\theta}{1+\cos\theta} = \dfrac{1-\cos\theta}{\sin\theta}$.

1. **Multiply the numerator and denominator of the fraction on the left by the conjugate of the denominator.**

$$\frac{\sin\theta}{1+\cos\theta} \cdot \frac{1-\cos\theta}{1-\cos\theta} = \frac{1-\cos\theta}{\sin\theta}$$

2. **Multiply the two denominators together, but leave the numerator in factored form.**

$$\frac{\sin\theta(1-\cos\theta)}{1-\cos^2\theta} = \frac{1-\cos\theta}{\sin\theta}$$

3. **Replace the denominator on the left with its equivalent by using the Pythagorean identity.**

$$\frac{\sin\theta(1-\cos\theta)}{\sin^2\theta} = \frac{1-\cos\theta}{\sin\theta}$$

4. **Reduce the fraction on the left.**

$$\frac{\cancel{\sin\theta}(1-\cos\theta)}{\sin^2\theta} = \frac{1-\cos\theta}{\sin\theta}$$

$$\frac{1-\cos\theta}{\sin\theta} = \frac{1-\cos\theta}{\sin\theta}$$

Squaring both sides

One special case of working on both sides of an identity at the same time is to square both sides. Your biggest clue as to when to use this technique is usually when one side or the other has a radical. This method is also good to use when you're solving some types of trig equations. Squaring both sides has two benefits: It gets rid of radicals, and it often creates terms that can be part of one of the Pythagorean identities. The Pythagorean identities have wonderful substitutions.

This first example has only one radical, and it's on the right side. Solve the identity $\dfrac{1-\cot x}{\csc x} = \sqrt{1-2\sin x \cos x}$.

1. **Square both sides of the identity.**

   Be sure to expand the squared binomial on the left correctly.

   $$\left(\frac{1-\cot x}{\csc x}\right)^2 = \left(\sqrt{1-2\sin x \cos x}\right)^2$$

   $$\frac{(1-\cot x)^2}{\csc^2 x} = 1-2\sin x \cos x$$

   $$\frac{1-2\cot x+\cot^2 x}{\csc^2 x} = 1-2\sin x \cos x$$

2. **Rearrange the terms in the numerator.**

   $$\frac{1+\cot^2 x-2\cot x}{\csc^2 x} = 1-2\sin x \cos x$$

3. **Replace $1 + \cot^2 x$ with its equivalent by using the Pythagorean identity.**

   $$\frac{\csc^2 x-2\cot x}{\csc^2 x} = 1-2\sin x \cos x$$

4. **Split up the fraction by writing each term in the numerator over the denominator.**

   $$\frac{\csc^2 x}{\csc^2 x} - \frac{2\cot x}{\csc^2 x} = 1-2\sin x \cos x$$

5. **Simplify the first term. Rewrite the numerator and denominator in the second term by using the ratio and reciprocal identities.**

   $$1 - \frac{2\dfrac{\cos x}{\sin x}}{\dfrac{1}{\sin^2 x}} = 1-2\sin x \cos x$$

6. **Simplify the complex fraction by flipping the denominator and multiplying it by the numerator.**

   $$1 - 2\frac{\cos x}{\cancel{\sin x}} \cdot \frac{\sin^2 x}{1} = 1-2\sin x \cos x$$

   $$1-2\cos x \sin x = 1-2\sin x \cos x$$

Identifying With the Operations

Identities that have sums, differences, multiple angles, and half-angles have a suggested procedure just staring at you, because having all the functions in terms of the same angle — not twice one or the sum of the other two — is best. You just decide which angle form you want everything to be in and then apply whatever identity the terms in the equation are equal to — substitute in the equivalence of the identity — and proceed from there.

Adding it up

Sum and difference identities usually involve two different angles and then a third combined angle. To prove the identity, you need to get rid of that third angle. The first example involves a sum of two different angles.

The equation $\dfrac{\cos(x+y)}{\cos x \cos y} = 1 - \tan x \tan y$ uses the angles x and y. Get rid of

the angle sum, $x + y$, by applying the appropriate identity, which contains just angle x and angle y.

1. **Replace the cosine of the sum of the two angles with its identity.**

    $$\frac{\cos x \cos y - \sin x \sin y}{\cos x \cos y} = 1 - \tan x \tan y$$

2. **Break up the fraction by putting each term in the numerator over the denominator.**

    $$\frac{\cos x \cos y}{\cos x \cos y} - \frac{\sin x \sin y}{\cos x \cos y} = 1 - \tan x \tan y$$

3. **Reduce the first fraction. Rewrite the second fraction as the product of two fractions. Then replace the two fractions in that product by using the ratio identity.**

    $$1 - \frac{\sin x}{\cos x} \cdot \frac{\sin y}{\cos y} = 1 - \tan x \tan y$$
    $$1 - \tan x \tan y = 1 - \tan x \tan y$$

The next example shows an identity for three times an angle: $\sin 3\theta = 3 \sin\theta - 4 \sin^2 \theta$.

1. **Replace the sine of 3θ with the sine of the sum of θ and 2θ to create the identity for the sum of two angles using the right side of the equation above.**

    $$\sin 3\theta = \sin(\theta + 2\theta)$$
    $$\sin(\theta + 2\theta) = 3 \sin\theta - 4 \sin^2 \theta$$

2. **Apply the angle-sum identity for sine on the left.**

   $\sin\theta\cos 2\theta + \cos\theta\sin 2\theta = 3\sin\theta - 4\sin^2\theta$

3. **Now replace cos 2θ and sin 2θ by using the double-angle identities.**

   You have two double-angle identities to choose from for cos 2θ. You choose the one involving the square of sine, because you see that same term on the right side of the equation.

   $\sin\theta(1 - 2\sin^2\theta) + \cos\theta(2\sin\theta\cos\theta) = 3\sin\theta - 4\sin^2\theta$

4. **Multiply through on the left side.**

   $\sin\theta - 2\sin^3\theta + 2\sin\theta\cos^2\theta = 3\sin\theta - 4\sin^2\theta$

5. **Replace cos² θ with its equivalent by using the Pythagorean identity. Then simplify the terms.**

   $$\sin\theta - 2\sin^3\theta + 2\sin\theta\left(1 - \sin^2\theta\right) = 3\sin\theta - 4\sin^2\theta$$

   $$\sin\theta - 2\sin^3\theta + 2\sin\theta - 2\sin^3\theta = 3\sin\theta - 4\sin^2\theta$$

   $$3\sin\theta - 4\sin^2\theta = 3\sin\theta - 4\sin^2\theta$$

What difference does it make?

Using functions involving the difference between two angle measures has many of the same features as those with sums, so I added a couple of twists to the examples in this section. The first identity uses the tangent of the difference of angles, $\tan(\alpha - \beta) = \dfrac{\tan\alpha - \tan\beta}{1 + \tan\alpha\tan\beta}$.

Now, solve the identity $\cot(x - 45°) = \dfrac{1 + \tan x}{\tan x - 1}$, which uses the tangent difference identity and incorporates some function values for a 45-degree angle.

1. **Rewrite the cotangent of the difference by using the reciprocal identity, because the cotangent doesn't have a standard difference identity.**

   $$\cot(x - 45°) = \frac{1}{\tan(x - 45°)}$$

2. **Now replace the tangent of the difference with its identity.**

   $$\frac{1}{\tan(x - 45°)} = \frac{1}{\dfrac{\tan x - \tan 45°}{1 + \tan x\tan 45°}}$$

3. **To simplify the complex fraction on the right, flip the denominator and multiply it by the numerator, which is 1.**

$$\frac{1}{\tan(x-45°)} = \frac{1+\tan x \tan 45°}{\tan x - \tan 45°}$$

4. **The value of the tangent of 45 degrees is 1, so replace all those terms with 1.**

$$\frac{1}{\tan(x-45°)} = \frac{1+\tan x(1)}{\tan x - 1}$$

$$\frac{1}{\tan(x-45°)} = \frac{1+\tan x}{\tan x - 1}$$

5. **Now rewrite the left side in its original form.**

$$\cot(x-45°) = \frac{1+\tan x}{\tan x - 1}$$

The next example proves that tan x is equal to itself. Yes, I know that seems a bit bizarre — obviously, it must be true. Discovering the technique involved is what makes going through the steps of this identity worth the effort. The trick here is to write tan x as the tangent of the difference of the angles $2x$ and x.

1. **Write the difference identity for tangent.**

$$\tan x = \tan x$$

$$\tan x = \tan(2x - x)$$

$$= \frac{\tan 2x - \tan x}{1 + \tan 2x \tan x}$$

2. **Replace the two terms tan 2x with the double-angle identity.**

$$\tan x = \frac{\dfrac{2\tan x}{1-\tan^2 x} - \tan x}{1 + \dfrac{2\tan x}{1-\tan^2 x} \cdot \tan x}$$

3. **Get rid of the complex fraction by multiplying every term in the numerator and denominator by 1 − tan² x.**

$$\tan x = \frac{\left(\dfrac{2\tan x}{1-\tan^2 x} - \tan x\right)(1-\tan^2 x)}{\left(1 + \dfrac{2\tan x}{1-\tan^2 x} \cdot \tan x\right)(1-\tan^2 x)}$$

$$\tan x = \frac{\left(\dfrac{2\tan x}{1-\tan^2 x}\right)(1-\tan^2 x) - \tan x(1-\tan^2 x)}{1(1-\tan^2 x) + \left(\dfrac{2\tan x}{1-\tan^2 x} \cdot \tan x\right)(1-\tan^2 x)}$$

4. **Simplify the fractions.**

$$\tan x = \frac{\left(\frac{2\tan x}{1-\tan^2 x}\right)(1-\tan^2 x) - \tan x(1-\tan^2 x)}{1(1-\tan^2 x) + \left(\frac{2\tan x}{1-\tan^2 x} \cdot \tan x\right)(1-\tan^2 x)}$$

$$= \frac{2\tan x - \tan x(1-\tan^2 x)}{1(1-\tan^2 x) + (2\tan x \cdot \tan x)}$$

5. **Multiply through the parentheses.**

$$\tan x = \frac{2\tan x - \tan x + \tan^3 x}{1-\tan^2 x + 2\tan^2 x}$$

6. **Combine the like terms in the numerator and denominator.**

$$\tan x = \frac{\tan x + \tan^3 x}{1+\tan^2 x}$$

7. **Factor the numerator. Then reduce the fraction.**

$$\tan x = \frac{\tan x(1+\tan^2 x)}{1+\tan^2 x} = \tan x$$

Multiplying your fun

The only special challenge involved in dealing with identities when using the multiple-angle formulas is in deciding which version of cos 2θ to use or whether to incorporate sums of angles or double angles. Here are some examples that illustrate these situations.

Solve the identity $\frac{\sin 2\theta + \cos 2\theta}{\sin\theta\cos\theta} = 2 + \cot\theta - \tan\theta$. You'll have to make a decision as to whether to use double-angle identities or sum identities, $\sin(\theta + \theta)$ and $\cos(\theta + \theta)$.

1. **The choice is double angle: apply the double-angle identities for sin 2θ and cos 2θ.**

$$\frac{2\sin\theta\cos\theta + \cos^2\theta - \sin^2\theta}{\sin\theta\cos\theta} = 2 + \cot\theta - \tan\theta$$

Choosing the formula for the sine isn't a problem, because you have only one to choose from. The cosine, however, takes some observations. Because you have both sine and cosine in the denominator, you don't want to use the double-angle identities for cosine that have a 1 in them.

2. Split up the fraction, writing each term in the numerator over the denominator.

$$\frac{2\sin\theta\cos\theta}{\sin\theta\cos\theta} + \frac{\cos^2\theta}{\sin\theta\cos\theta} - \frac{\sin^2\theta}{\sin\theta\cos\theta} = 2 + \cot\theta - \tan\theta$$

3. Reduce the fractions.

$$\frac{2\cancel{\sin\theta}\,\cancel{\cos\theta}}{\cancel{\sin\theta}\,\cancel{\cos\theta}} + \frac{\cos^2\theta}{\sin\theta\,\cancel{\cos\theta}} - \frac{\sin^2\theta}{\cancel{\sin\theta}\,\cos\theta} = 2 + \cot\theta - \tan\theta$$

$$2 + \frac{\cos\theta}{\sin\theta} - \frac{\sin\theta}{\cos\theta} = 2 + \cot\theta - \tan\theta$$

4. Replace the fractions with their equivalents by using the ratio identities.

$$2 + \cot\theta - \tan\theta = 2 + \cot\theta - \tan\theta$$

Finding an identity for an angle with a multiple greater than 2 requires that you decide whether to use a sum identity or a double-angle identity. The best approach isn't always clear. Sometimes the other terms in the equation give you a hint. Often, you just flip a coin. Solve the identity $\sin 4x = 4 \sin x \cos^3 x - 4 \sin^3 x \cos x$.

1. Rewrite sin 4x as sin (2 · 2x).

$$\sin (2 \cdot 2x) = 4 \sin x \cos^3 x - 4 \sin^3 x \cos x$$

2. Insert this new angle into the double-angle identity.

$$2 \sin 2x \cos 2x = 4 \sin x \cos^3 x - 4 \sin^3 x \cos x$$

3. Replace sin 2x and cos 2x with the double-angle identities, choosing the best cosine identity for the situation.

In this case, because you have a difference of terms involving third-degree powers of sine and cosine, the cosine identity involving second-degree powers of sine and cosine seems to be a good choice.

$$2(2 \sin x \cos x)(\cos^2 x - \sin^2 x) = 4 \sin x \cos^3 x - 4 \sin^3 x \cos x$$

4. Multiply through.

$$2(2\sin x\cos x)\left(\cos^2 x - \sin^2 x\right) = 4\sin x\cos^3 x - 4\sin^3 x\cos x$$

$$4\sin x\cos x\left(\cos^2 x - \sin^2 x\right) = 4\sin x\cos^3 x - 4\sin^3 x\cos x$$

$$4\sin x\cos^3 x - 4\sin^3 x\cos x = 4\sin x\cos^3 x - 4\sin^3 x\cos x$$

Halving fun, wish you were here

The last type of identity that you can incorporate into solving an identity or doing identity problems is the half-angle identity. This identity actually comes in very handy in calculus — not by changing from a half-angle identity to angles that are larger, but by changing from larger angles to half-angle identities. The examples in this section show some of the possibilities.

Solve the identity $\tan\frac{\theta}{2} = \csc\theta - \cot\theta$ by using the half-angle identity for tangent. Neither side looks very promising for solving the identity until you notice that you have two different angles — one half the size of the other. You need to apply the half-angle identity to get everything in terms of the same angle.

1. **Substitute in the identity for the half-angle of tangent.**

   You have two different versions to choose between,

   $\tan\frac{\theta}{2} = \frac{\sin\theta}{1+\cos\theta} = \frac{1-\cos\theta}{\sin\theta}$. The easier one to work with is the one with two terms in the numerator: $\frac{1-\cos\theta}{\sin\theta} = \csc\theta - \cot\theta$.

2. **Split up the fraction on the left by putting each term in the numerator over the denominator.**

   $\frac{1}{\sin\theta} - \frac{\cos\theta}{\sin\theta} = \csc\theta - \cot\theta$

3. **Replace the two terms on the left with their reciprocal and ratio identities.**

   $\csc\theta - \cot\theta = \csc\theta - \cot\theta$

This last example incorporates the half-angle of the tangent, as well as the half-angle of a reciprocal function. Solve the identity $\sin\theta\sec^2\frac{\theta}{2} = 2\tan\frac{\theta}{2}$. You see here that working on both sides of the equation is necessary.

1. **Use the reciprocal of the half-angle identity for cosine to replace the half-angle of secant.**

   Hold off on deciding which version of the tangent's half-angle formula to use until you see what you need.

   $$\sin\theta\left(\frac{1}{\cos\frac{\theta}{2}}\right)^2 = 2\tan\frac{\theta}{2}$$

   $$\sin\theta\left(\frac{1}{\sqrt{\frac{1+\cos\theta}{2}}}\right)^2 = 2\tan\frac{\theta}{2}$$

2. **Flip the fraction in the parentheses and square the radical, leaving the terms with no radical.**

$$\sin\theta\left(\frac{2}{1+\cos\theta}\right)=2\tan\frac{\theta}{2}$$

3. **Choose the half-angle tangent identity that matches what you have on the left and simplify.**

$$\sin\theta\left(\frac{2}{1+\cos\theta}\right)=2\left(\frac{\sin\theta}{1+\cos\theta}\right)$$

$$\frac{2\sin\theta}{1+\cos\theta}=\frac{2\sin\theta}{1+\cos\theta}$$

Part IV
Equations and Applications

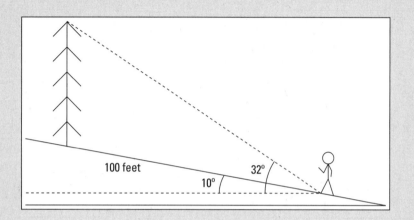

In this part...

- ✔ Become acquainted with inverse trig functions.
- ✔ Identify the domains and ranges of the inverse trig functions.
- ✔ Recognize the pairings of the quadrants used by each inverse function.
- ✔ Solve trig equations using identities and inverse functions.
- ✔ Write expressions to include infinitely many answers.
- ✔ Find the areas of triangles using trig functions in the formulas.

Chapter 15

Investigating Inverse Trig Functions

As thrilling and fulfilling as the original six trig functions are, they just aren't complete without their inverses. An *inverse trig function* behaves like the inverse of any other type of function — it undoes what the original function did. In mathematics, functions can have inverses if they're one-to-one, meaning each output value occurs only once. This whole inverse idea is going to take some fast talking when it comes to trig functions, because they keep repeating values over and over as angles are formed with every full rotation of the circle — so you're going to wonder how these functions and inverses can be one-to-one. If you need a refresher on basic inverse functions, flip on back to Chapter 3 for the lowdown on them and how you determine one.

Writing It Right

You use inverse trig functions to solve equations such as $\sin x = \frac{1}{2}$

or sec $x = -2$, or tan $2x = 1$. In typical algebra equations, you can solve for the value of x by dividing each side of the equation by the coefficient or by adding the same thing to each side, and so on. But you can't do that with the function $\sin x = \frac{1}{2}$.

Would it make sense to divide each side by *sine?* "Out, out thou sine!" Here's what you'd get: $\dfrac{\sin x}{\sin} = \dfrac{\frac{1}{2}}{\sin}$, $x = \dfrac{\frac{1}{2}}{\sin}$. Goodness, no! That's silly.

Using the notation

Using inverses allows you to determine the value of x in a trig equation. To find the inverse of an equation such as $\sin x = \frac{1}{2}$, means to complete the following statement: "x is equal to the angle whose sine is equal to $\frac{1}{2}$." In trig speak, you write this statement as $x = \sin^{-1}\left(\frac{1}{2}\right)$. This standard notation involves putting a –1 in the superscript position immediately following the function name. Here are some more examples of trig equations with their corresponding inverses and the translation.

Function	Inverse	What It Means
$\sec x = -2$	$x = \sec^{-1}(-2)$	x is the angle whose secant is –2
$\tan 2x = 1$	$2x = \tan^{-1}(1)$	$2x$ is the angle whose tangent is 1
$\cos \theta = 0$	$\theta = \cos^{-1}(0)$	θ is the angle whose cosine is 0
$\csc \alpha = -1$	$\alpha = \csc^{-1}(-1)$	α is the angle whose cosecant is –1

Interpreting the exponent

You've undoubtedly seen and used the exponent –1 in math expressions before now. But that exponent does a different kind of job for inverse trig functions and relations. The notation for an inverse trig relation such as $\tan^{-1} x$ means that you want an *inverse* for the expression, not the *reciprocal.* If you really want the reciprocal of tangent, $\dfrac{1}{\tan x}$, then you have to write with parentheses $(\tan x)^{-1}$. Of course, the reciprocal of tangent is cotangent. The –1 exponent is where the exponential notation for trig functions makes a big exception.

When raising trig functions to a power, $\sin^2 x = (\sin x)^2$ and $\cos^4 x = (\cos x)^4$, but $\tan^{-1} x$ means the inverse function, not raising $\tan x$ to the –1 power.

Alternating the notation

Inverses of trig functions have an alternate notation that avoids the confusion over what the –1 superscript means: the *arc* name. Another way of saying $\sin^{-1} x$ is arcsin x. The inverse cosine is $\cos^{-1} x$, or arccos x. The other inverse functions are arctan x, arccsc x, arcsec x, and arccot x. This notation is longer to write out and is sometimes awkward to write, so the original superscript notation is often preferable. You'll see inverse functions written both ways, though.

Distinguishing between the few and the many

Technically, an inverse function is supposed to have only one answer. (Part of the definition of an inverse is that the function and inverse are one-to-one.) Each input has one output, and each output has one input. To accommodate all the practical uses of trig inverses, you have a way around this rule. You can designate whether you want one answer or many answers by using either the inverse *function* or the inverse *relation*. A relation is a bit looser than a function; it allows more than one input to have the same output. To differentiate between these two entities, I will use a capital letter for the name of a function and a lowercase letter for the corresponding relation.

Trig Functions	**Trig Relations**
$Sin^{-1} x$ or Arcsin x	$sin^{-1} x$ or arcsin x
$Cos^{-1} x$ or Arccos x	$cos^{-1} x$ or arccos x
$Tan^{-1} x$ or Arctan x	$tan^{-1} x$ or arctan x
$Cot^{-1} x$ or Arccot x	$cot^{-1} x$ or arccot x
$Sec^{-1} x$ or Arcsec x	$sec^{-1} x$ or arcsec x
$Csc^{-1} x$ or Arccsc x	$csc^{-1} x$ or arccsc x

If you evaluate the function $Sin^{-1}\left(\frac{1}{2}\right)$, the result is 30 degrees (or $\frac{\pi}{6}$ radians). Just one answer exists, which is called the *principal value* of the inverse. But if you write $sin^{-1}\left(\frac{1}{2}\right)$, then the result can be 30 degrees, 150 degrees, 390 degrees, 510 degrees, and so on (or $\frac{\pi}{6}, \frac{5\pi}{6}, \frac{13\pi}{6}, \frac{17\pi}{6}, \frac{25\pi}{6}, \frac{29\pi}{6}, ...$). It all depends on the situation — what you want at the time. Do you want just the principal value, or do you want multiple values? Or you may want a bunch of values within one full rotation — from 0 to 360 degrees.

When you want lots and lots of angles or answers, listing them all can be tedious. In fact, listing every possible solution may not even be doable. Rather than making a list, you can give a *rule*, which allows you to define an angle with all its full-rotation multiples — the angles with the same terminal side.

Let *n* represent any integer {... , –3, –2, –1, 0, 1, 2, 3, ...}. Using the *n* as a multiplier, you can write a long list of angles more efficiently. Instead of saying $x = 30, 150, 390, 510, 750, 870, ...$, divide the list into two groups: $x = 30; 390;$ 750; 1,110; ... ; and $x = 150; 510; 870; 1,230; ...$; and then use the two rules that follow:

$$x = 30 + 360n \text{ or } x = 150 + 360n$$

Also, in radians, instead of saying $x = \dfrac{\pi}{6}, \dfrac{13\pi}{6}, \dfrac{25\pi}{6}, \cdots$ or $x = \dfrac{5\pi}{6}, \dfrac{17\pi}{6}, \dfrac{29\pi}{6}, \cdots$, use these two rules: $x = \dfrac{\pi}{6} + 2\pi n$ or $x = \dfrac{5\pi}{6} + 2\pi n$.

Here's an example showing how to write all the angles that have a cosine equal to $\dfrac{1}{2}$. The steps involve solving the inverse relation, not just finding the principal value for the function. Solve for the values that satisfy $x = \cos^{-1}\left(\dfrac{1}{2}\right)$.

1. **List several solutions in both degrees and radians.**

   $\cos^{-1}\left(\dfrac{1}{2}\right) = 60°, 300°, 420°, 660°, 780°, 1020°, \ldots$

   $\cos^{-1}\left(\dfrac{1}{2}\right) = \dfrac{\pi}{3}, \dfrac{5\pi}{3}, \dfrac{7\pi}{3}, \dfrac{11\pi}{3}, \dfrac{13\pi}{3}, \dfrac{17\pi}{3}, \ldots$

2. **Write the answers in degrees by using just the first two angles plus multiples of 360.**

   $\cos^{-1}\left(\dfrac{1}{2}\right) = 60° + 360°n$ or $\cos^{-1}\left(\dfrac{1}{2}\right) = 300° + 360°n$

3. **Write the answers in radians by using the first two angles plus multiples of 2π.**

   $\cos^{-1}\left(\dfrac{1}{2}\right) = \dfrac{\pi}{3} + 2\pi n$ or $\cos^{-1}\left(\dfrac{1}{2}\right) = \dfrac{5\pi}{3} + 2\pi n$

Writing all the possible angles for inverse tangent is a bit easier than for sine or cosine. The tangent is positive in the first and third quadrants, which are cattycorner from one another (half a full rotation). Because of this fact, the angles that have the same function values are 180 degrees apart, and you can use nice multiples of 180 degrees or π to name all the answers. This isn't the case with sine and cosine, though. The angles with the same function values are in quadrants that are adjacent to one another, so you have to use two separate rules — both with multiples of 360 degrees — to name all the answers.

Here's how to write all the angles that have a tangent equal to $-\dfrac{\sqrt{3}}{3}$. Solve for values that satisfy $x = \tan^{-1}\left(-\dfrac{\sqrt{3}}{3}\right)$.

1. **List several answers in both degrees and radians.**

   $\tan^{-1}\left(-\dfrac{\sqrt{3}}{3}\right) = 150°, 330°, 510°, 690°, \ldots$

   $\tan^{-1}\left(-\dfrac{\sqrt{3}}{3}\right) = \dfrac{5\pi}{6}, \dfrac{11\pi}{6}, \dfrac{17\pi}{6}, \dfrac{23\pi}{6}, \ldots$

2. **Write the answers in degrees by using multiples of 180.**

$$\tan^{-1}\left(-\frac{\sqrt{3}}{3}\right) = 150° + 180°n$$

3. **Write the answers in radians by using multiples of π.**

$$\tan^{-1}\left(-\frac{\sqrt{3}}{3}\right) = \frac{5\pi}{6} + \pi n$$

Determining Domain and Range of Inverse Trig Functions

A function that has an inverse has exactly one output (belonging to the *range*) for every input (belonging to the *domain*), and vice versa. To keep inverse trig functions consistent with this definition, you have to designate ranges for them that will take care of all the possible input values and not have any duplication. The output values of the inverse trig functions are all angles — in either degrees or radians — and they're the answer to the question, "Which angle gives me this number?" In general, the output angles for the individual inverse functions are paired up as angles in Quadrants I and II or angles in Quadrants I and IV. The quadrants are selected this way for the inverse trig functions because the pairs are adjacent quadrants, allowing for both positive and negative entries. The notation for these inverse functions uses capital letters (see the preceding section).

Inverse sine function

The domain for $\text{Sin}^{-1} x$, or Arcsin x, is from –1 to 1. In mathematical notation, the domain or input values, the x's, fit into the expression $-1 \le x \le 1$, because no matter what angle measure you put into the sine function, the output of the function lies between –1 and 1, including those two numbers. The range, or output, for $\text{Sin}^{-1} x$ is all angles from –90 to 90 degrees or, in radians, $-\frac{\pi}{2}$ to $\frac{\pi}{2}$. Because the output of the inverse sine function is some angle θ, you write this range as $-90° \le \theta \le 90°$ or $-\frac{\pi}{2} \le \theta \le \frac{\pi}{2}$. The outputs of the inverse sine function are angles in the adjacent Quadrants I and IV, because the sine is positive in the first quadrant and negative in the second quadrant. Those angles cover all the possible input values — their function values represent all the numbers from –1 to 1.

Inverse cosine function

The domain for $\text{Cos}^{-1} x$, or Arccos x, is from –1 to 1, just like the inverse sine function. So the x (or input) values are $-1 \leq x \leq 1$. The range for $\text{Cos}^{-1} x$ consists of all angles from 0 to 180 degrees or, in radians, 0 to π. Because the output of the inverse cosine is some angle θ, you write these expressions for the range as $0° \leq \theta \leq 180°$ or $0 \leq \theta \leq \pi$. The outputs are angles in the adjacent Quadrants I and II, because the cosine is positive in the first quadrant and negative in the second quadrant. Those angles cover all the possible input values for the function.

Inverse tangent function

The domain for $\text{Tan}^{-1} x$, or Arctan x, is all real numbers — numbers from $-\infty$ to ∞. This is because the output of the tangent function, this function's inverse, includes all numbers. The range, or output, of $\text{Tan}^{-1} x$ is angles between –90 and 90 degrees or, in radians, between $-\frac{\pi}{2}$ and $\frac{\pi}{2}$. One important note is that the range doesn't include those beginning and ending angles; the tangent function isn't defined for –90 or 90 degrees. The range of $\text{Tan}^{-1} x$ includes all the angles in the adjacent Quadrants I and IV, except for the two angles with terminal sides on the y-axis.

Inverse cotangent function

The domain of $\text{Cot}^{-1} x$, or Arccot x, is the same as that of the inverse tangent function. The domain includes all real numbers. The range, though, is different — it includes all angles between 0 and 180 degrees (between 0 and π). So any angle in Quadrants I and II is included in the range, except for those with terminal sides on the x-axis. Those two angles aren't in the domain of the cotangent function (see Chapter 8), so they aren't in the range of the inverse.

Inverse secant function

The domain of $\text{Sec}^{-1} x$, or Arcsec x, consists of all the numbers from 1 on up plus all the numbers from –1 on down. Letting x be the input, you write this expression for the domain as $x \geq 1$ or $x \leq -1$. In other words, the domain includes all the numbers from $-\infty$ to ∞, except for the numbers between -1 and 1. The range of $\text{Sec}^{-1} x$ is all the angles from 0 to 180 degrees (from 0 to π) — meaning all angles in Quadrants I and II, with the exception of 90 degrees, or $\frac{\pi}{2}$.

Inverse cosecant function

The domain of $\text{Csc}^{-1} x$, or $\text{Arccsc } x$, is the same as that for the inverse secant function, all the numbers from 1 on up plus all the numbers from -1 on down. The range is different, though — it includes all angles from -90 to 90 degrees or, in radians, from $-\frac{\pi}{2}$ to $\frac{\pi}{2}$. In short, the range is all the angles in Quadrants I and IV, with the exception of 0 degrees, or 0 radians.

Summarizing domain and range

Sometimes, looking at a chart or summary of the domains and ranges of the inverse trig functions is more informative than reading about them. Take a look at Table 15-1. You should notice some patterns — some similarities and differences. The ranges of three of the functions are in Quadrants I and II, and the other three are in Quadrants I and IV. The reciprocals sine and cosecant use the same quadrants. So do the reciprocals cosine and secant. The tangent and cotangent don't use the same quadrants, though.

Table 15-1	**Domains and Ranges of the Inverse Trig Functions**		
Inverse Trig Function	**Domain**	**Range**	**Quadrants in Range**
$\text{Sin}^{-1} x$	$-1 \leq x \leq 1$	$-90° \leq \theta \leq 90°$	I and IV
$\text{Cos}^{-1} x$	$-1 \leq x \leq 1$	$0° \leq \theta \leq 180°$	I and II
$\text{Tan}^{-1} x$	$-\infty < x < \infty$	$-90° < \theta < 90°$	I and IV
$\text{Cot}^{-1} x$	$-\infty < x < \infty$	$0° < \theta < 180°$	I and II
$\text{Sec}^{-1} x$	$x \geq 1$ or $x \leq -1$	$0° \leq \theta \leq 180°, \theta \neq 90°$	I and II
$\text{Csc}^{-1} x$	$x \geq 1$ or $x \leq -1$	$-90° \leq \theta \leq 90°, \theta \neq 0°$	I and IV

Triangle dissection paradox

Look at the two triangles in the following figure, made up of the same four pieces. The pieces are the same size in each triangle. They're sitting on the same grid, appearing to take up the same space, but one has a hole or gap. How can that be?

This situation happens because the hypotenuse of the main outer triangle isn't really a straight line. The slopes of the sides of the

two triangles (the hypotenuses) making up that larger hypotenuse aren't the same; the hypotenuse of the top triangle is slightly bent in, and the hypotenuse of the bottom triangle is slightly bent out. Of course, you're not supposed to be able to notice this bend, but you can check it by laying a ruler or straightedge along the hypotenuse. You might call this figure an optical illusion.

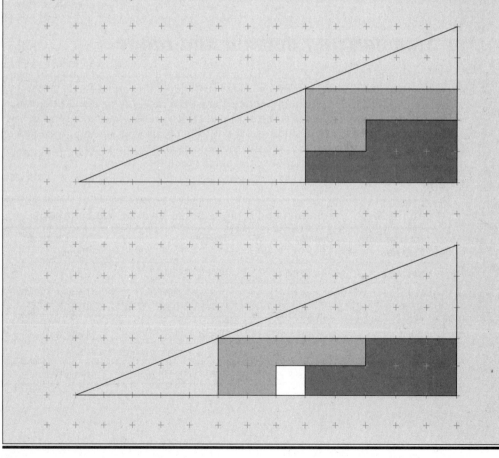

Chapter 16

Making Inverse Trig Work for You

*I*n Chapter 15, I introduce you to the six inverse trig functions. As with many introductions to something new, it may take a while to place the name with the face (or in this case, properties). This chapter, on the other hand, takes you deeper into the world of inverses and shows how the inverses of trig functions work. You'll also see why you'd even want to bother learning the names with their good traits or bad.

Working with Inverses

The easiest way to work with inverse trig functions is to have a chart handy with the exact values of the functions, which you can find in the Appendix. When angles other than the most common or popular are involved, you can either use a table such as the one in the Appendix or get out your handy-dandy scientific calculator.

This first example on evaluating an inverse uses the exact value from a chart. Find $\cos^{-1}\left(-\frac{\sqrt{2}}{2}\right)$.

1. **Determine the reference angle that you need by using the absolute value of the input.**

   The value $\frac{\sqrt{2}}{2}$ is the cosine of 45 degrees or $\frac{\pi}{4}$ radians.

2. **Use the sign of the input to determine the correct quadrant.**

   Because $-\frac{\sqrt{2}}{2}$ is negative, and of the two quadrants for the range (refer to Table 15-1), the cosine is negative in QII, the answer is an angle in QII whose reference angle is 45 degrees.

3. **Determine the correct angle measure.**

   The angle in standard position in QII whose reference angle is 45 degrees is $180 - 45 = 135$ or $\pi - \frac{\pi}{4} = \frac{3\pi}{4}$. So $\cos^{-1}\left(-\frac{\sqrt{2}}{2}\right) = 135°$ or $\frac{3\pi}{4}$. (Refer to Chapter 8 for more on reference angles.)

The next example involves inverse cotangent. Find $\cot^{-1}\left(\sqrt{3}\right)$.

1. **Determine the reference angle that you need.**

   The value $\sqrt{3}$ is the cotangent of 30 degrees or $\frac{\pi}{6}$.

   You either have that memorized or found that in the Appendix.

2. **Use the sign of the input to determine the correct quadrant.**

   Because the cotangent is positive in QI, then the answer is the angle in QI whose reference angle is 30 degrees or $\frac{\pi}{6}$.

3. **Determine the correct angle measure.**

   All angles in QI are the same as their reference angles, so $\cot^{-1}\left(-\sqrt{3}\right) = 30°$ or $\frac{\pi}{6}$.

The problems that you encounter won't always involve nice numbers from the most common angles. When you're faced with a nasty little decimal value, you may have to use a table. In this next example, you start off with a decimal value; an answer to the nearer the proper response. The decimal in the following example is rounded off to three decimal places. To do these problems, you find the closest answer.

Find Arctan(–3.732).

1. **Determine the reference angle that you need.**

   Use the Appendix, and you can see that the value 3.732 corresponds to the tangent of a 75-degree angle. This angle is the closest in whole degrees to having a tangent of 3.732.

2. **Use the sign of the input to determine the correct quadrant.**

   Because –3.732 is negative, the answer is an angle in QIV whose reference angle is 75 degrees.

3. **Determine the correct angle measure.**

In QIV, a reference angle of 75 degrees has a measure of either –75 degrees or its positive equivalent (same terminal side), 285 degrees. Because the range of Arctan(x) includes only angles between –90° and 90°, we choose –75 degrees. So, Arctan(–3.732) = Tan^{-1}(–3.732) = –75°.

If you want your answer in radians, use the formula from Chapter 5, $\frac{\theta°}{180°} = \frac{\theta^R}{\pi}$.

You use the formula to find either the negative angle or positive angle and then do some algebra to find the other. Finding the radian equivalent of –75°,

$$\frac{-75°}{180°} = \frac{\theta^R}{\pi}$$

$$\frac{-5°}{12} = \frac{\theta^R}{\pi}$$

$$\frac{-5°}{12}\pi = \theta^R$$

Getting Friendly with Your Calculator

Scientific calculators are wonderful tools — they make life easier and improve the quality (correctness) of the results. In most instances, computing inverse trig functions with a calculator is quick and easy. You need to be aware of a couple of pitfalls, though.

If you've already trekked to the store and left empty-handed and bewildered by the array of graphing calculators, your best bet is to hold your head high, confidently head back to the store, and pick up a model by Texas Instruments, Sharp, or Hewlett-Packard — those models seem to be the most popular. (For what it's worth, I'm a Texas Instruments gal — especially when it comes to their graphing calculators.)

Changing the mode

Scientific calculators are very accommodating — they give you results in either degrees or radians, depending on which *mode* you set them in. This feature is great, but it trips up even the best mathematicians from time to time. Each calculator is different, but they always have either a single button or a multiple-button sequence that switches from radians to degrees and back again — sort of like a toggle switch. Some calculators even have a legend at the top or bottom of the screen that tells you whether you're in degree or radian mode, perhaps as obvious as an *R* or *D*. Problems tend to arise when you use your calculator for more than one task. Perhaps you're studying both trigonometry and physics, and one calls for degree mode while the other calls for radian mode. Just be aware, and you won't get caught.

Interpreting notation on the calculator

Calculator *notation,* or the mumbo jumbo on the buttons, is somewhat tricky. Even though the –1 superscript indicates an inverse trig function when written in a book or on your paper, you can't use the –1 button to find the value of an inverse trig function on a calculator. On scientific calculators, the –1 or x^{-1} button means to find the reciprocal of a number. Look under the *2nd functions,* which are different functions or operations written above the buttons, for the inverse trig functions. They're usually above the original sine, cosine, and tangent buttons. Some calculators have a button labeled "2nd." Others use alternate colors — usually yellow, red, or green — to denote the second use of the button.

Even when you find the inverse functions, you'll notice that they're only for the three primary trig functions. The calculator doesn't show any for cosecant, secant, or cotangent. So, where are they? First, I discuss how to use the three buttons available; then I tell you how to calculate the other inverses.

Using the inverse function button

To explain this button, I use an example. Here's how you find $\sin^{-1}\left(\frac{1}{2}\right)$ in degrees using a scientific calculator.

1. **Decide whether you want your answer in radians or degrees.**

   For this example, use the mode menu or whatever method your calculator uses to change the mode to Degrees.

2. **Enter the problem as given.**

   The following are the typical keystrokes: 2nd sin .5 Enter . The result is 30, meaning 30 degrees.

Now looks what happens when you decide to find $\sin^{-1}\left(\frac{1}{2}\right)$ in radians using that same calculator.

1. **You decide that you want your answer in radians.**

   Use the mode menu or whatever method your calculator uses to change the mode to Radians.

2. **Enter the problem as given.**

   The following are the same keystrokes: 2nd sin .5 Enter . The result is 0.5235987756 on my calculator. This is in radians. If a radian is about 57 degrees, and this decimal is over half, then the answer is somewhere near 30 degrees — which you know from the previous problem. But what do you do if you haven't done a previous problem?

3. **Divide π by the decimal value.**

   Don't skimp on the decimal places. Use your calculator and enter: π ÷ .5235987756 Enter or, on my calculator, I use: π ÷ Ans Enter , because I have the decimal sitting there from the previous computation.

   The result is 6; the 6 is the denominator under π, which represents the radian measure equivalent to 0.5235987756. So, the radian answer in terms of π is $\frac{\pi}{6}$.

Calculating the inverse of a reciprocal function

To determine the inverse of a reciprocal function, such as $\text{Cot}^{-1}(2)$ or $\text{Sec}^{-1}(-1)$, you have to change the problem back to the function's reciprocal — one of the three basic functions — and then use the appropriate inverse button.

When changing to the function's reciprocal, you flip the input with that function, too. For example, $\text{Cot}^{-1}(2)$ becomes $\tan^{-1}\left(\frac{1}{2}\right)$. You change $\sec^{-1}\left(\frac{2}{\sqrt{3}}\right)$ to $\cos^{-1}\left(\frac{\sqrt{3}}{2}\right)$, $\text{Csc}^{-1}(1)$ to $\text{Sin}^{-1}(1)$, and so on.

For example, find the value of $\text{Sec}^{-1}(-1.1547)$. This time find the answer in degrees.

1. **Rewrite the function in terms of its reciprocal.**

   Find the reciprocal of –1.1547 using your calculator.

   –1.1547 x^{-1} Enter

   The result is –0.8660258076. Just use the first four decimal places in your work.

   $\text{Sec}^{-1}(-1.1547) = \text{Arccos}(-0.8660)$

2. **Enter the problem.**

   2nd cos –.8660 Enter

The calculator give you 149.9970891. This is essentially an angle of 150 degrees. All the rounding of decimals inserts a bit of an error. You can check by finding the cosine of a 150-degree angle using your calculator and then finding the reciprocal, to see if you have the correct answer. Also, the exact value of $\cos(150)$ is $-\frac{\sqrt{3}}{2}$, which as a decimal value of 0.8660254038. Look familiar?

Working around the inverse cotangent

The other big pitfall you encounter when using a scientific calculator involves the inverse cotangent. The inverse tangent, $\text{Tan}^{-1}x$, has its range in QI and QIV, but $\text{Cot}^{-1}x$ has its range in QI and QII. If you want $\cot^{-1}\left(-\sqrt{3}\right)$, for example, and you use $\tan^{-1}\left(-\frac{1}{\sqrt{3}}\right)$ and your calculator, you get an answer in the fourth quadrant. You have to be aware that this quadrant isn't correct; you got it because you changed functions from cotangent to tangent so you could use the calculator. This is still the best way to do the problem. Just use the answer from the calculator and determine the corresponding angle in QII. Here's an example:

Find $\cot^{-1}\left(-\sqrt{3}\right)$ in degrees.

1. **Set the mode to Degrees.**

2. **Change the function and value to their reciprocals.**

   $\cot^{-1}\left(-\sqrt{3}\right)$ becomes $\tan^{-1}\left(-\frac{1}{\sqrt{3}}\right)$.

3. **Find the value of the inverse function by using a calculator.**

   Enter [2nd][tan][(−)]1/√3 [Enter]. On some calculators, parentheses will automatically pop up for you to enter the tangent value inside them. If they don't, then you should insert parentheses around the fraction yourself. The result is –30 degrees. Note that this angle is in QIV, and you want QII for the inverse cotangent.

4. **Find the angle in QII that has the same reference angle.**

   The angle in QII with a 30-degree reference angle is an angle of 150 degrees. So $\cot^{-1}\left(-\sqrt{3}\right)=150°$.

Multiplying the Input

Multiple-angle functions are those such as $\sin 2\theta$, $\cos 3x$, $\tan 6\beta$, and so on. When considering inverse relations (which give multiple answers) for these angles, the multiplier usually indicates how many more answers that problem has (it's how many more *times*) as compared to a similar problem without a multiplier. For example, the equation $\sin\theta = \frac{\sqrt{3}}{2}$ has two different answers if you consider all the angles between 0 and 360 degrees: θ equals 60 and 120 degrees. But if you change the equation to $\sin 2\theta = \frac{\sqrt{3}}{2}$, you get twice as many, or four, answers between 0 and 360 degrees: θ equals 30, 60, 210, and 240 degrees. These angles are all within one rotation, but putting them into the original equation and multiplying by 2 gives angles with the same terminal side as the angles within one rotation.

Here are some examples to show you how this multiplication works and how to find the answers. First, I show how I got the answers for $\sin 2\theta = \frac{\sqrt{3}}{2}$.

1. **Write the inverse equation.**

$$2\theta = \sin^{-1}\left(\frac{\sqrt{3}}{2}\right)$$

2. **List all the angles in two rotations, $0° \leq \theta \leq 720°$, that have that sine, and set them equal to 2θ.**

$2\theta = 60°, 120°, 420°, 480°$

The second two angles are each 360 more than the first two.

3. **Divide each of the terms on both sides of the equation by 2 to solve for θ.**

$$\frac{2\theta}{2} = \frac{60°}{2}, \frac{120°}{2}, \frac{420°}{2}, \frac{480°}{2}$$
$$\theta = 30°, 60°, 210°, 240°$$

Notice how all the solutions for θ are between 0 and 360 degrees — just as asked.

Now solve $\cos 3x = -\frac{\sqrt{2}}{2}$ for any x such that $0 \leq x \leq 2\pi$. Note that this interval indicates radian measures.

1. **Write the inverse equation.**

$$3x = \cos^{-1}\left(-\frac{\sqrt{2}}{2}\right)$$

2. **List all the angles in three rotations, $0 \le x \le 6\pi$, that have that cosine, and set them equal to $3x$.**

$$3x = \frac{3\pi}{4}, \frac{5\pi}{4}, \frac{11\pi}{4}, \frac{13\pi}{4}, \frac{19\pi}{4}, \frac{21\pi}{4}$$

The second two angles are just 2π greater than the first two, and the last two are 2π greater than the second two. Just change 2π to $\frac{8\pi}{4}$ and add the fractions.

3. **Multiply all the terms on both sides by $\frac{1}{3}$ to solve for x.**

$$\frac{1}{3} \cdot 3x = \frac{1}{3} \cdot \frac{\cancel{3}\pi}{4}, \frac{1}{3} \cdot \frac{5\pi}{4}, \frac{1}{3} \cdot \frac{11\pi}{4}, \frac{1}{3} \cdot \frac{13\pi}{4}, \frac{1}{3} \cdot \frac{19\pi}{4}, \frac{1}{\cancel{3}} \cdot \frac{\cancel{21}^{7}\pi}{4}$$

$$x = \frac{\pi}{4}, \frac{5\pi}{12}, \frac{11\pi}{12}, \frac{13\pi}{12}, \frac{19\pi}{12}, \frac{7\pi}{4}$$

This result shows the big advantage of radians — the numbers don't get as big as they do with degrees. The disadvantage may be having so many fractions.

Solving Some Mixed Problems

When working with inverse trig functions, it's always more convenient when the numbers you're working with are the results of applying one of the trig functions to a common angle measure. The exact values of the functions of those more popular angles are easy to remember and work with in problems. When the angle isn't a common one, though, you need a calculator or table. Not a big deal, just not as pleasant.

By using inverse trig functions, you can solve some interesting problems, where you never even need to know what the angle measure actually is. You just need to know a function value, a quadrant, and a few trig identities.

For example, you can find $\cos\left(\sin^{-1}\left(-\frac{12}{13} \right) \right)$, which says to "find the cosine of an angle whose sine is equal to $\frac{12}{13}$." You don't need to know the angle measure to solve this problem, but you do need to know the quadrant that the terminal side lies in, because otherwise, two different angles can be correct answers. The sine is positive in QI and QII, so this problem could involve an angle in either of those quadrants, but cosine isn't positive in both of those quadrants. Consider the following example.

Find $\cos\left(\sin^{-1}\left(-\frac{12}{13}\right)\right)$ if the terminal side of the angle is in QII.

1. **Use the Pythagorean identity to find the numerical value of the cosine of the angle.**

   Put the value in for $\sin\theta$, get the cosine term alone, and then take the square root of both sides:

   $$\sin^2\theta + \cos^2\theta = 1$$

   $$\left(\frac{12}{13}\right)^2 + \cos^2\theta = 1$$

   $$\cos^2\theta = 1 - \frac{144}{169} = \frac{25}{169}$$

   $$\cos\theta = \pm\sqrt{\frac{25}{169}} = \pm\frac{5}{13}$$

2. **Choose the sign of the answer.**

   Because the angle's terminal side is in QII, and the cosine is negative there, the answer is $\cos\left(\sin^{-1}\left(-\frac{12}{13}\right)\right) = -\frac{5}{13}$.

The quadrant isn't a mystery in a problem that uses the inverse trig function. The previous example includes information on the terminal side of the angle — in which quadrant it lies. When an inverse function is involved, the quadrant is spelled out for you by the range of the function involved. You just use the assigned quadrants.

For example, to find $\tan\left(\cos^{-1}\left(-\frac{11}{61}\right)\right)$, you can assume that the angle has its terminal side in QII, because the inverse cosine function is negative in that quadrant.

1. **Use the reciprocal identity and reciprocal of the number to find the secant.**

   The problem involves the angle whose cosine is $-\frac{11}{61}$. I call that unknown angle θ and rewrite the expression in terms of the cosine of θ with that measure. I write the expression this way in order to change from an inverse trig function to a trig function so I can use the identity.

   If $\cos\theta = -\frac{11}{61}$, then $\sec\theta = -\frac{61}{11}$.

2. **Use the Pythagorean identity to solve for the tangent.**

$$\tan^2\theta + 1 = \sec^2\theta = \left(-\frac{61}{11}\right)^2 = \frac{3{,}721}{121}$$

$$\tan^2\theta = \frac{3{,}721}{121} - 1 = \frac{3{,}600}{121}$$

$$\tan^2\theta = \pm\sqrt{\frac{3{,}600}{121}}\, 1 = \pm\frac{60}{11}$$

3. **Choose the sign of the answer.**

Because the terminal side is in QII and the tangent is negative in that quadrant, $\tan\left(\cos^{-1}\left(-\frac{11}{61}\right)\right) = -\frac{60}{11}$.

Chapter 17

Solving Trig Equations

. . . .

In This Chapter

▶ Solving equations within limits

▶ Expanding the pool of answers for equations

▶ Incorporating algebra techniques

▶ Getting creative with identities

▶ Solving multiple-angle equations

▶ Letting a graphing calculator do the dirty work

. . . .

Solving equations involving trigonometric expressions takes a pinch of this, a dab of that, a gentle stirring, and just the right temperature. No, this book isn't for a cooking class, but solving these equations requires the proper preparation and some skill — just like a successful dish.

Trig equations aren't identities. An identity is true for any angle in the domain of the function involved. A trig *equation* is true for some specific angles or input — if the equation has a solution at all.

Some trig equations require factoring skills from algebra or even the quadratic formula. Successfully solving most trig equations involves incorporating trig identities at the proper time. All the equations require knowledge of the function values and how inverse trig functions work (so head on back to Chapters 9, 15 and 16 if you need a refresher). For the equation-solving enthusiasts, this chapter is where all the concepts come together for maximum fun and challenge.

The methods and techniques that you see in this chapter are those that people most frequently use to solve trig equations. A few more ways exist, but they don't come up as often. Also, you usually have more than one way of solving a particular trig equation. Your goal should always be to do it as quickly and efficiently as possible, but don't be alarmed if you seem to take the long way around. Sometimes the more circuitous route just seems to make more sense. If a particular method works for you — in other words, you get the right answer — go for it!

Generating Simple Solutions

The simplest type of trig equation is the one that you can immediately rewrite as an inverse in order to determine the solutions. Some examples of these types of equations include $\cos x = 1$, $2 \sin x + 1 = 0$, and $\cot x - \sqrt{3} = 0$. Here's how to solve them.

To solve $\cos x = 1$:

1. **Rewrite the equation as an inverse function equation.**

   $x = \cos^{-1}(1)$

2. **List the solutions for values of x when $0 \leq x < 360°$.**

   $x = 0°$

   The only time that the cosine is equal to 1 is when the angle, or input, is 0 degrees.

3. **List all the solutions in general.**

   $x = 0° + 360°n$

Steps 2 and 3 illustrate the different ways that you can write the answers: either as a few within a certain interval, or as all that are possible, with a rule to describe them.

Now solve $2 \sin x + 1 = 0$ only for values of x such that $0 \leq x < 2\pi$:

1. **Rewrite the equation as an inverse function equation.**

   First, subtract 1 from each side; then divide each side by 2.

   $2 \sin x = -1$

   $\sin x = -\dfrac{1}{2}$

   $x = \sin^{-1}\left(-\dfrac{1}{2}\right)$

2. **List the solutions. Use the chart in the Appendix to find the angles that work.**

   $x = \dfrac{7\pi}{6}, \dfrac{11\pi}{6}$

This last example involves a reciprocal function. Your best bet is to begin by using a reciprocal identity and changing the problem.

Solve the equation $\cot x - \sqrt{3} = 0$ for all the values of x, in radians, that satisfy it:

1. **Solve for the trig function by adding the radical value to each side.**

   $\cot x = \sqrt{3}$

2. **Use the reciprocal identity and the reciprocal of the number to change to the tangent function, and then multiply both parts of the fraction by the denominator to get rid of the radical.**

   $\tan x = \dfrac{1}{\sqrt{3}} = \dfrac{\sqrt{3}}{3}$

3. **Rewrite the equation as an inverse function equation.**

   $x = \tan^{-1}\left(\dfrac{\sqrt{3}}{3}\right)$

4. **Write the general statements that give all the solutions.**

   $x = \dfrac{\pi}{6} + \pi n$

   $x = \dfrac{\pi}{6}, \dfrac{7\pi}{6}, \dfrac{13\pi}{6}, \dfrac{19\pi}{6}, \dots$

   These statements mean that all the angles you find by adding or sub-tracting multiples of π will provide solutions for this equation.

Factoring In the Solutions

The same type of factoring that algebra uses is a great help in solving trig equations. The only trick is to recognize that instead of just x's, y's, or other single-letter variables, trig variables such as $\sin x$ or $\sec y$ exist. You need the whole $\sin x$ or $\sec y$ when working with the variables. You can't factor out an x or a sec alone. Look for the patterns and apply the factoring techniques. Here's a list of the basic factoring patterns.

Factoring binomials:

- ✔ **Greatest common factor:** $ab \pm cb = b(a \pm c)$
- ✔ **Difference of squares:** $a^2 - b^2 = (a + b)(a - b)$
- ✔ **Sum or difference of cubes:** $a^3 + b^3 = (a + b)(a^2 - ab + b^2)$

  $a^3 - b^3 = (a - b)(a^2 + ab + b^2)$

Factoring trinomials:

✔ **Greatest common factor:** $ax^2 + ax + ac = a(x^2 + x + c)$

✔ **Un-FOIL:** $abx^2 + (ad + bc)x + cd = (ax + c)(bx + d)$

Factoring by grouping:

$$abxy + adx + bcy + cd = ax(by + d) + c(by + d) = (ax + c)(by + d)$$

Finding a greatest common factor

The trig equations that require finding a greatest common factor (GCF), factoring it out, and then solving the equation could look like these two equations: $2 \sin x \cos x - \sin x = 0$ or $\cos x \tan x = \sqrt{3} \cos x$. I solve both of these equations in this section.

Solve $2 \sin x \cos x - \sin x = 0$ for all the values of x such that $0 \leq x < 360°$.

1. **Factor out sin x from each of the two terms.**

   $\sin x(2 \cos x - 1) = 0$

2. **Set the two different factors equal to 0.**

   $\sin x = 0$ or $2 \cos x - 1 = 0$

3. **Solve for the values of x that satisfy each equation. Use the table in the Appendix.**

   If $\sin x = 0$, then $x = \sin^{-1}(0) = 0°, 180°$.

   If $2 \cos x - 1 = 0$, $2 \cos x = 1$, $\cos x = \frac{1}{2}$, then $x = \cos^{-1}\left(\frac{1}{2}\right) = 60°, 300°$.

   All these values are solutions for the original equation. The complete list is $x = 0°, 60°, 180°, 300°$.

Now solve $\cos x \tan x = \sqrt{3} \cos x$ for all the possible values in degrees.

1. **Move the term on the right to the left by subtracting it from each side.**

   $\cos x \tan x - \sqrt{3} \cos x = 0$

2. **Factor out the cos x from each term.**

   $\cos x \left(\tan x - \sqrt{3}\right) = 0$

   You don't want to divide each side by cos x, because you'll lose a solution if you do.

3. **Set the two different factors equal to 0.**

   $\cos x = 0$ or $\tan x - \sqrt{3} = 0$

4. **Solve for the values of x that satisfy both equations.**

   If $\cos x = 0$, then $x = \cos^{-1}(0) = 90°, 270°, \ldots$ or $90° + 180°n$.

   If $\tan x - \sqrt{3} = 0, \tan x = \sqrt{3}$, then $x = \tan^{-1}\left(\sqrt{3}\right) = 60°, 240°$ or $60° + 180°n$.

   So the solutions are all of the form $x = 90° + 180°n$ or $x = 60° + 180°n$.

Factoring quadratics

Quadratic equations are nice to work with because, when they don't factor, you can solve them by using the quadratic formula (see the "Using the Quadratic Formula" section later in this chapter). The types of quadratic trig equations that you can factor are those like $\tan^2 x = \tan x$, $4\cos^2 x - 3 = 0$, $2\sin^2 x + 5\sin x - 3 = 0$, or $\csc^2 x + \csc x - 2 = 0$. Notice that each equation has the telltale trig function raised to the second degree. I show you how to handle them in the following examples.

The first two examples have just two terms. The first has two variable terms, and the other has just one variable term. In the first example, you put both terms on the left and then factor out the variable or trig term.

Solve $\tan^2 x = \tan x$ for the values of x such that $0 \le x < 2\pi$.

1. **Move the term $\tan x$ on the right to the left by subtracting it from both sides.**

   $\tan^2 x - \tan x = 0$

   Don't divide through by $\tan x$. You'll lose solutions.

2. **Factor out $\tan x$.**

   $\tan x(\tan x - 1) = 0$

3. **Set each of the two factors equal to 0.**

   $\tan x = 0$ or $\tan x - 1 = 0$

4. **Solve for the values of x that satisfy both equations.**

   If $\tan x = 0$, then $x = \tan^{-1}(0) = 0, \pi$.

   If $\tan x - 1 = 0, \tan x = 1$, then $x = \tan^{-1}(1) = \frac{\pi}{4}, \frac{5\pi}{4}$.

   The four solutions are $x = 0, \frac{\pi}{4}, \pi$, and $\frac{5\pi}{4}$.

In this next example, the binomial doesn't factor easily as the difference of two squares, because the 3 isn't a perfect square, and you have to use a radical in the factorization. A nice, efficient way to solve this equation is to move the 3 to the right and take the square root of each side.

Solve for all the possible solutions of $4\cos^2 x - 3 = 0$ in degrees.

1. **Move the number to the right by adding 3 to each side.**

   $4\cos^2 x = 3$

2. **Take the square root of each side. Then solve for cos x by dividing each side by 2.**

   $$\sqrt{4\cos^2 x} = \pm\sqrt{3}$$
   $$2\cos x = \pm\sqrt{3}$$
   $$\cos x = \pm\frac{\sqrt{3}}{2}$$

3. **Solve the two equations for the values of x.**

   If $\cos x = \frac{\sqrt{3}}{2}$, then $x = \cos^{-1}\left(\frac{\sqrt{3}}{2}\right) = 30°, 330°$, with all the possible solutions being $x = 30° + 360°n$ or $330° + 360°n$.

   If $\cos x = -\frac{\sqrt{3}}{2}$, then $x = \cos^{-1}\left(-\frac{\sqrt{3}}{2}\right) = 150°, 210°$, with all the possible solutions being $x = 150° + 360°n$ or $210° + 360°n$.

   When you consider all the multiples of 360 degrees added to the four base angles, you find that this equation has lots and lots of solutions.

The next two examples involve using un-FOIL — a technique for determining which two binomials give you a particular quadratic trinomial. When the pattern in the trinomial is obscured, you may want to first substitute some other variable for the trig function to help figure out how you factor it. I do this when I solve $2\sin^2 x + 5\sin x - 3 = 0$ for all the values of x between 0 and 360 degrees.

1. **Replace each sin x with y.**

   $2y^2 + 5y - 3 = 0$

2. **Factor the trinomial as the product of two binomials.**

   $(2y - 1)(y + 3) = 0$

3. **Replace each y with sin x.**

   $(2\sin x - 1)(\sin x + 3) = 0$

4. Set each factor equal to 0.

$2 \sin x - 1 = 0$ or $\sin x + 3 = 0$

5. Solve the two equations for the values of x that satisfy them.

If $2 \sin x - 1 = 0$, $2 \sin x = 1$, $\sin x = \frac{1}{2}$, then $x = \sin^{-1}\left(\frac{1}{2}\right) = 30°, 150°$.

If $\sin x + 3 = 0$, $\sin x = -3$, then $x = \sin^{-1}(-3)$. This result is nonsense, because the sine function only produces values between -1 and 1 — so this factor doesn't produce any solutions. Go to Chapter 9 for information on the range of the sine function.

The only two solutions are 30 and 150 degrees.

This next example factors fairly easily, but it involves a reciprocal function. Solve $\csc^2 x + \csc x - 2 = 0$ for any angles between 0 and 2π radians.

1. Factor the quadratic trinomial into the product of two binomials.

$(\csc x + 2)(\csc x - 1) = 0$

2. Set each factor equal to 0.

$\csc x + 2 = 0$ or $\csc x - 1 = 0$

3. Solve the two equations for the values of x that satisfy them.

If $\csc x + 2 = 0$, $\csc x = -2$, then $x = \csc^{-1}(-2) = \frac{7\pi}{6}, \frac{11\pi}{6}$.

If $\csc x - 1 = 0$, $\csc x = 1$, then $x = \csc^{-1}(1) = \frac{\pi}{2}$.

An alternate way of dealing with these two binomial equations is to change them by using the reciprocal identity and writing the reciprocal of the number. For the first equation, you'd change from cosecant to sine: $\csc x + 2 = 0$, $\csc x = -2$, $\sin x = -\frac{1}{2}$. Do the same for the second equation: $\csc x - 1 = 0$, $\csc x = 1$, $\sin x = 1$. You'd then solve the inverse equations (and get the same answers).

Increasing the degrees in factoring

Factoring quadratics is a breeze — well, I guess it gets a bit windy at times. Factoring equations with higher degrees can get a bit nasty if you don't have a nice situation such as just two terms or a quadratic-like equation. You may have the possibility of factoring by grouping, and I cover that method in the next section. In this section, the problems that I have in mind are those like $2 \sin^3 x = \sin x$ or $2 \cos^4 x - 9 \cos^2 x + 4 = 0$.

The first equation has just two terms, so you can factor it by finding a greatest common factor. Solve $2 \sin^3 x = \sin x$ for all the possible angles in degrees.

1. **Move the term on the right to the left by subtracting it from each side.**

   $2 \sin^3 x - \sin x = 0$

2. **Factor out sin x.**

   $\sin x (2 \sin^2 x - 1) = 0$

3. **Set each factor equal to 0.**

   $\sin x = 0$ or $2 \sin^2 x - 1 = 0$

4. **Solve the two equations for the values of x that satisfy them.**

   If $\sin x = 0$, then $x = \sin^{-1}(0) = 0°, 180°, \ldots$ or $0° + 180°n$.

   If $2 \sin^2 x - 1 = 0$, $2 \sin^2 x = 1$, $\sin^2 x = \frac{1}{2}$, then you end up with a quadratic equation.

5. **Take the square root of both sides of the equation and solve for x.**

   Multiply both parts of the fraction by the denominator to get the radical out of the denominator.

   $$\sqrt{\sin^2 x} = \pm\sqrt{\frac{1}{2}}$$

   $$\sin x = \pm\frac{\sqrt{2}}{2}$$

   Now, considering both solutions:

   If $\sin x = \frac{\sqrt{2}}{2}$, then $x = \sin^{-1}\left(\frac{\sqrt{2}}{2}\right) = 45°, 135°, \ldots$ or $45° + 360°n, 135° + 360°n$.

   If $\sin x = -\frac{\sqrt{2}}{2}$, then $x = \sin^{-1}\left(-\frac{\sqrt{2}}{2}\right) = 225°, 315°, \ldots$ or $225° + 360°n, 315° + 360°n$.

This third-degree trig equation has a whole slew of answers:

$x = 180°n$

$x = 45° + 360°n$

$x = 135° + 360°n$

$x = 225° + 360°n$

$x = 315° + 360°n$

TIP

You can combine these last four equations for x, the ones that begin with multiples of 45 degrees, to read $x = 45° + 90°n$. This single equation generates all the same angles as the last four statements combined. How do you know you can simplify this way? Because the angles of 45, 135, 225, and 315 degrees are all 90 degrees apart in value. By starting with the 45 and adding 90 over and over, you get all the listed angles, as well as the infinite number of their multiples.

The next example is a fourth-degree equation, but this one is quadratic-like, meaning that it factors like a quadratic trinomial with two binomial factors. This problem has the possibility of having a great number of solutions — or none. Solve $2\cos^4 x - 9\cos^2 x + 4 = 0$ for the solutions that are between 0 and 2π.

1. **Factor the trinomial as the product of two binomials.**

   $(2\cos^2 x - 1)(\cos^2 x - 4) = 0$

2. **Set each factor equal to 0.**

   $2\cos^2 x - 1 = 0$ or $\cos^2 x - 4 = 0$

3. **Solve for the function in each equation by getting the cosine terms alone on one side of the equation.**

   $2\cos^2 x - 1 = 0$

   $2\cos^2 x = 1$ and $\cos^2 x - 4 = 0$

   $\cos^2 x = \dfrac{1}{2}$ $\cos^2 x = 4$

4. **Take the square root of each side of each equation.**

   $\sqrt{\cos^2 x} = \pm\sqrt{\dfrac{1}{2}}$

   $\cos x = \pm\dfrac{\sqrt{2}}{2}$ and $\sqrt{\cos^2 x} = \pm\sqrt{4}$

   $\cos x = \pm 2$

5. **Solve for the values of x that satisfy the equations.**

   If $\cos x = \dfrac{\sqrt{2}}{2}$, then $x = \cos^{-1}\left(\dfrac{\sqrt{2}}{2}\right) = \dfrac{\pi}{4}, \dfrac{7\pi}{4}$.

   If $\cos x = -\dfrac{\sqrt{2}}{2}$, then $x = \cos^{-1}\left(-\dfrac{\sqrt{2}}{2}\right) = \dfrac{3\pi}{4}, \dfrac{5\pi}{4}$.

   If $\cos x = \pm 2$, then you have a problem — that equation doesn't compute! The cosine function results in values between –1 and 1. (Find out more about the range of cosine in Chapter 9.) This factor doesn't give any new solutions to the original problem.

Factoring by grouping

The process of factoring by grouping works in very special cases; one case is when the original equation is the result of multiplying two binomials together that have some unrelated terms in them. You usually can apply this type of factoring when you're facing an even number of terms and can find common factors in different groups of them. The types of equations that you can solve by using grouping may look like $4 \sin x \cos x - 2 \sin x - 2 \cos x + 1 = 0$ or $\sin^2 x \sec x + 2 \sin^2 x = \sec x + 2$. In the first equation, the first two terms have an obvious common factor, $2 \sin x$. The second two have no common factor other than 1, but to make grouping work, factor out -1.

Solve $4 \sin x \cos x - 2 \sin x - 2 \cos x + 1 = 0$ for all the possible answers between 0 and 2π.

1. **Factor $2 \sin x$ out of the first two terms and -1 out of the second two.**

   $2 \sin x(2 \cos x - 1) - 1(2 \cos x - 1) = 0$

   Now you have two terms on the left, each with a factor of $2 \cos x - 1$.

2. **Factor that common factor out of the two terms.**

   $(2 \cos x - 1)(2 \sin x - 1) = 0$

3. **Set the two factors equal to 0.**

   $$2\cos x - 1 = 0 \qquad 2\sin x - 1 = 0$$
   $$2\cos x = 1 \qquad\quad 2\sin x = 1$$
   $$\cos x = \frac{1}{2} \qquad\quad \sin x = \frac{1}{2}$$

4. **Solve for the values of x that satisfy the equation.**

   If $\cos x = \frac{1}{2}$, then $x = \cos^{-1}\left(\frac{1}{2}\right) = \frac{\pi}{3}, \frac{5\pi}{3}$.

   If $\sin x = \frac{1}{2}$, then $x = \sin^{-1}\left(\frac{1}{2}\right) = \frac{\pi}{6}, \frac{5\pi}{6}$.

This next example of grouping requires that you begin by moving the two terms on the right to the left. Another twist is that one of the resulting factors turns out to be a quadratic. How can math be much more fun than this?

Solve $\sin^2 x \sec x + 2 \sin^2 x = \sec x + 2$ for all the angles between 0 and 360 degrees.

1. **Move the terms on the right to the left by subtracting them from both sides.**

   $\sin^2 x \sec x + 2 \sin^2 x - \sec x - 2 = 0$

2. **Factor $\sin^2 x$ out of the first two terms and -1 out of the second two.**

   $\sin^2 x (\sec x + 2) - 1(\sec x + 2) = 0$

3. **Now factor $\sec x + 2$ out of the two terms.**

   $(\sec x + 2)(\sin^2 x - 1) = 0$

4. **Set the two factors equal to 0.**

   $\sec x + 2 = 0$, $\sec x = -2$

   $\sin^2 x - 1 = 0$, $\sin^2 x = 1$, $\sin x = \pm 1$

   when you take the square root of both sides.

5. **Solve for the values of x that satisfy the equations.**

   If $\sec x = -2$, then $x = \sec^{-1}(-2) = 120°, 240°$.

   If $\sin x = 1$, then $x = \sin^{-1}(1) = 90°$.

   If $\sin x = -1$, then $x = \sin^{-1}(-1) = 270°$.

Using the Quadratic Formula

When quadratic equations factor, life is good. When they don't, you can still survive, thanks to that wonderful quadratic formula. In case you've forgotten the exact formula, here it is.

The quadratic formula says that if you have a quadratic equation in the form $ax^2 + bx + c = 0$, then its solutions are $x = \dfrac{-b \pm \sqrt{b^2 - 4ac}}{2a}$.

In trig, a trig function replaces the x or variable part of the quadratic formula. For example, find the solution of $\sin^2 x - 4 \sin x - 1 = 0$ for all angles between 0 and 360 degrees. Instead of just x's, the variable terms are $\sin x$'s.

1. **Identify the values of the a, b, and c in the formula.**

   The values are $a = 1$, $b = -4$, and $c = -1$.

2. **Fill in the quadratic formula with these values and simplify.**

$$\sin x = \frac{-(-4) \pm \sqrt{(-4)^2 - 4(1)(-1)}}{2(1)}$$

$$= \frac{4 \pm \sqrt{16 + 4}}{2} = \frac{4 \pm \sqrt{20}}{2}$$

$$= \frac{4 \pm 2\sqrt{5}}{2} = 2 \pm \sqrt{5}$$

3. **Find approximate values for sin x from the solved form.**

Using a scientific calculator, $2 \pm \sqrt{5} \approx 2 \pm 2.236$. So, $\sin x$ is either about 4.236 or –0.236.

4. **Use a table of values to find approximate angles with these sines.**

If $\sin x = 4.236$, you get an impossible result. The value of the sine ranges from –1 to 1, so $\sin x$ can't have this value.

If $\sin x = -0.236$, then $x = \sin^{-1}(-0.236) \approx -14°$ or $346°$. These are the same angle. First, you write it as a negative angle and then as its positive equivalent.

Another angle satisfies this equation, too. The other negative angle that has a 14-degree reference angle is the third-quadrant angle of 194 degrees. Refer to Chapter 8 for more on reference angles.

Incorporating Identities

Some trig equations contain more than one trig function. Others have mixtures of multiple angles and single angles with the same variable. Some examples of such equations include $3 \cos^2 x = \sin^2 x$, $2 \sec x = \tan x + \cot x$, and $\cos 2x + \cos x + 1 = 0$. To get these equations into a more-manageable form so that you can use factoring or one of the other methods in this chapter to solve them, you call upon identities to substitute some or all of the terms (for more on basic trig identities, see Chapter 11). For example, to solve $3 \cos^2 x = \sin^2 x$ for all the angles between 0 and 2π, apply the Pythagorean identity.

1. **Replace sin² x with its equivalent from the Pythagorean identity, sin² x + cos² x = 1 or sin² x = 1 – cos² x.**

$3 \cos^2 x = 1 - \cos^2 x$

2. **Add $\cos^2 x$ to each side and simplify by dividing.**

$$4\cos^2 x = 1$$

$$\cos^2 x = \frac{1}{4}$$

3. **Take the square root of each side.**

$$\sqrt{\cos^2 x} = \pm\sqrt{\frac{1}{4}}$$

$$\cos x = \pm\frac{1}{2}$$

4. **Solve for the values of x that satisfy the equation.**

If $\cos x = \frac{1}{2}$, then $x = \cos^{-1}\left(\frac{1}{2}\right) = \frac{\pi}{3}, \frac{5\pi}{3}$.

If $\cos x = -\frac{1}{2}$, then $x = \cos^{-1}\left(-\frac{1}{2}\right) = \frac{2\pi}{3}, \frac{4\pi}{3}$.

In this next example, you begin with three different trig functions. A good tactic is to replace each function by using either a ratio identity or a reciprocal identity. Using these identities creates fractions, and fractions require common denominators. By the way, having fractions in trig equations is *good*, because the products that result from multiplying and making equivalent fractions are usually parts of identities that you can then substitute in to make the expression much simpler. Solve $2 \sec x = \tan x + \cot x$ for all the possible solutions in degrees.

1. **Replace each term with its respective reciprocal or ratio identity.**

$$2\left(\frac{1}{\cos x}\right) = \frac{\sin x}{\cos x} + \frac{\cos x}{\sin x}$$

$$\frac{2}{\cos x} = \frac{\sin x}{\cos x} + \frac{\cos x}{\sin x}$$

2. **Rewrite the fractions with the common denominator $\sin x \cos x$.**

Multiply each term by a fraction that equals 1, with either sine or cosine in both the numerator and denominator.

$$\frac{2}{\cos x} \cdot \frac{\sin x}{\sin x} = \frac{\sin x}{\cos x} \cdot \frac{\sin x}{\sin x} + \frac{\cos x}{\sin x} \cdot \frac{\cos x}{\cos x}$$

$$\frac{2\sin x}{\sin x \cos x} = \frac{\sin^2 x}{\sin x \cos x} + \frac{\cos^2 x}{\sin x \cos x}$$

3. **Add the two fractions on the right. Then, using the Pythagorean identity, replace the new numerator with 1.**

$$\frac{2\sin x}{\sin x \cos x} = \frac{\sin^2 x + \cos^2 x}{\sin x \cos x}$$

$$\frac{2\sin x}{\sin x \cos x} = \frac{1}{\sin x \cos x}$$

4. **Set the equation equal to 0 by subtracting the right term from each side.**

Perform the subtraction to create a single fraction.

$$\frac{2\sin x - 1}{\sin x \cos x} = 0$$

5. **Now set the numerator equal to 0.**

$$2\sin x - 1 = 0 \text{ or } 2\sin x = 1, \sin x = \frac{1}{2}$$

If the numerator is equal to 0, then the whole fraction is equal to 0. The denominator can't equal 0 — such a number doesn't exist.

6. **Solve for the values of x that satisfy the original equation.**

$$x = \sin^{-1}\left(\frac{1}{2}\right) = 30°, 150°, \ldots$$

$$x = 30° + 360°n \text{ or } x = 150° + 360°n$$

In the next example, two different angles are in play. One angle is twice the size of the other, so you use a double-angle identity to reduce the terms to only one angle. The trick is to choose the correct version of the cosine double-angle identity.

Solve $\cos 2x + \cos x + 1 = 0$ for x between 0 and 2π.

1. **Replace $\cos 2x$ with $2\cos^2 x - 1$.**

$$2\cos^2 x - 1 + \cos x + 1 = 0$$

This version of the cosine double-angle identity is preferable because the other trig function in the equation already has a cosine in it.

2. **Simplify the equation. Then factor out $\cos x$.**

$$2\cos^2 x + \cos x = 0$$

$$\cos x (2\cos x + 1) = 0$$

3. **Set each factor equal to 0.**

$$\cos x = 0 \text{ or } 2\cos x + 1 = 0, 2\cos x = -1, \cos x = -\frac{1}{2}$$

4. **Solve for the values of x that satisfy the original equation.**

   If $\cos x = 0$, then $x = \cos^{-1}(0) = \dfrac{\pi}{2}, \dfrac{3\pi}{2}$.

   If $\cos x = -\dfrac{1}{2}$, then $x = \cos^{-1}\left(-\dfrac{1}{2}\right) = \dfrac{2\pi}{3}, \dfrac{4\pi}{3}$.

This last example looks deceptively simple. The catch is that you have to recognize a double-angle identity upfront and make a quick switch. This example is also a nice segue into the next section on equations with multiple-angle solutions. Solve $\sin x \cos x = \dfrac{1}{2}$ for all the solutions between 0 and 360 degrees.

1. **Use the sine double-angle identity to create a substitution for the expression on the left.**

   Starting out with the identity and multiplying each side by $\dfrac{1}{2}$, you get:
   $$\sin 2x = 2\sin x \cos x$$
   $$\frac{1}{2}\sin 2x = \sin x \cos x$$

2. **Replace the expression on the left of the original equation with its equivalent from the double-angle identity.**
   $$\sin x \cos x = \frac{1}{2}$$
   $$\frac{1}{2}\sin 2x = \frac{1}{2}$$

3. **Multiply each side of the equation by 2.**
   $$2 \cdot \frac{1}{2}\sin 2x = \frac{1}{2} \cdot 2$$
   $$\sin 2x = 1$$

4. **Rewrite the expression as an inverse function.**
   $$2x = \sin^{-1}(1)$$

   See Chapters 15 and 16 for more on inverse functions.

5. **Determine which angles within *two* rotations satisfy the expression.**
   $$2x = \sin^{-1}(1) = 90°, 450°$$

   You use two rotations because the coefficient of x is 2.

6. **Divide each term by 2.**
   $$\frac{2x}{2} = \frac{90°}{2}, \frac{450°}{2}$$
   $$x = 45°, 225°$$

   Notice that the resulting angles are between 0 and 360 degrees.

You can generalize the double-angle technique from the preceding example for other multiple-angle expressions.

Finding Multiple-Angle Solutions

Multiple-angle expressions are those where the angle measure is some multiple of a variable — for example, $2x$ or $3y$. In this section, I show you how to take these expressions apart and solve for all the additional solutions that are possible. Because the trig functions are *periodic* (meaning they repeat their patterns infinitely), the number of possibilities for solutions increases tremendously. The larger the multiplier, the more possible solutions.

When solving a trig equation of the form $ax = f^{-1}(x)$ where you want the solution to be all the angles within one complete rotation, write out all the solutions within the number of complete rotations that a represents. Then divide each angle measure by a.

Problems that lend themselves to this technique are those such as $2 \sin^2 5x = 1$ and $\cos\left(\frac{1}{2}x\right) = -\frac{\sqrt{3}}{2}$. In the first example, I solve $2 \sin^2 5x = 1$ for all the angles between 0 and 2π.

1. **Divide each side by 2; then take the square root of each side.**

   $$\sin^2 5x = \frac{1}{2}$$

   $$\sqrt{\sin^2 5x} = \pm\sqrt{\frac{1}{2}} = \pm\frac{\sqrt{2}}{2}$$

   $$\sin 5x = \pm\frac{\sqrt{2}}{2}$$

2. **Solve for $5x$, which represents the angles that satisfy the equation within one rotation.**

   If $\sin 5x = \frac{\sqrt{2}}{2}$, then $5x = \sin^{-1}\left(\frac{\sqrt{2}}{2}\right) = \frac{\pi}{4}, \frac{3\pi}{4}$.

   If $\sin 5x = -\frac{\sqrt{2}}{2}$, then $5x = \sin^{-1}\left(-\frac{\sqrt{2}}{2}\right) = \frac{5\pi}{4}, \frac{7\pi}{4}$.

3. **Extend the solutions to five rotations by adding 2π to each of the original angles four times.**

   $$5x = \frac{\pi}{4}, \frac{3\pi}{4}, \frac{9\pi}{4}, \frac{11\pi}{4}, \frac{17\pi}{4}, \frac{19\pi}{4}, \frac{25\pi}{4}, \frac{27\pi}{4}, \frac{33\pi}{4}, \frac{35\pi}{4}$$

   and

   $$5x = \frac{5\pi}{4}, \frac{7\pi}{4}, \frac{13\pi}{4}, \frac{15\pi}{4}, \frac{21\pi}{4}, \frac{23\pi}{4}, \frac{29\pi}{4}, \frac{31\pi}{4}, \frac{37\pi}{4}, \frac{39\pi}{4}$$

4. Divide all the terms by 5 and simplify.

$$x = \frac{\pi}{20}, \frac{3\pi}{20}, \frac{9\pi}{20}, \frac{11\pi}{20}, \frac{17\pi}{20}, \frac{19\pi}{20}, \frac{25\pi}{20}, \frac{27\pi}{20}, \frac{33\pi}{20}, \frac{35\pi}{20}$$

$$x = \frac{\pi}{20}, \frac{3\pi}{20}, \frac{9\pi}{20}, \frac{11\pi}{20}, \frac{17\pi}{20}, \frac{19\pi}{20}, \frac{5\pi}{4}, \frac{27\pi}{20}, \frac{33\pi}{20}, \frac{7\pi}{4}$$

and

$$x = \frac{5\pi}{20}, \frac{7\pi}{20}, \frac{13\pi}{20}, \frac{15\pi}{20}, \frac{21\pi}{20}, \frac{23\pi}{20}, \frac{29\pi}{20}, \frac{31\pi}{20}, \frac{37\pi}{20}, \frac{39\pi}{20}$$

$$x = \frac{\pi}{4}, \frac{7\pi}{20}, \frac{13\pi}{20}, \frac{3\pi}{4}, \frac{21\pi}{20}, \frac{23\pi}{20}, \frac{29\pi}{20}, \frac{31\pi}{20}, \frac{37\pi}{20}, \frac{39\pi}{20}$$

Notice that all 16 solutions are angles with measures less than 2π.

This next example has a proper-fraction multiplier rather than a multiplier greater than 1.

Solve $\cos\left(\frac{1}{2}x\right) = -\frac{\sqrt{3}}{2}$ for all the solutions between 0 and 360 degrees.

1. Rewrite the equation as an inverse trig equation.

$$\frac{1}{2}x = \cos^{-1}\left(-\frac{\sqrt{3}}{2}\right)$$

2. Determine which angles satisfy the inverse equation within one full rotation.

$$\frac{1}{2}x = \cos^{-1}\left(-\frac{\sqrt{3}}{2}\right) = 150°, 210°$$

3. Multiply all the terms by 2.

$$x = 300°, 420°$$

4. Throw out the second angle, because its measure is greater than 360 degrees.

The only solution is 300 degrees. When you replace the x in the original equation with this angle measure, you get a true statement.

Squaring Both Sides

When solving trig equations, you have so many choices to choose from as techniques to use for the solution. Many times, more than one method will work — although one method is usually quicker or easier than another. And then you'll come across a trig equation that defies your finest attempts. Two last-ditch efforts that you can use when solving trig equations are to square both sides of the equation or to multiply each term through by a trig function that you've carefully selected. I show you the first of these two methods here, and I show you the second method in the next section. Examples of equations that respond well to squaring both sides include $\sin x + \cos x = \sqrt{2}$ and $\cos x - \sqrt{3}\sin x = 1$.

Solve $\sin x + \cos x = \sqrt{2}$ for all the possible angles in degrees.

1. **Square both sides of the equation.**

   When squaring a binomial, be sure not to forget the middle term.
   $$\left(\sin x + \cos x\right)^2 = \left(\sqrt{2}\right)^2$$
   $$\sin^2 x + 2\sin x \cos x + \cos^2 x = 2$$

2. **Use the Pythagorean identity to replace $\sin^2 x + \cos^2 x$ with the number 1.**
   $$\sin^2 x + \cos^2 x + 2\sin x \cos x = 2$$
   $$1 + 2\sin x \cos x = 2$$

3. **Subtract 1 from each side. Then replace the expression on the left using the sine double-angle formula.**
   $$2\sin x \cos x = 1$$
   $$\sin 2x = 1$$

4. **Solve for the value of $2x$ by using the inverse function. Then write a few angle solutions to determine a pattern.**
   $$2x = \sin^{-1}(1) = 90°, \ 450°, \ 810°, \ ...$$

   Because you're supposed to find all the possible solutions, you're not bound by only two rotations.

5. **Divide every term by 2.**
   $$2x = 90°, 450°, 810°, ...$$
   $$x = 45°, 225°, 405°, ...$$

6. **Check for extraneous solutions.**

   Because squaring both sides of the equation loses information about the signs, you may have introduced extraneous solutions. Plug in the values of x that you found to check:

   $\sin\left(45°\right) + \cos\left(45°\right) = \sqrt{2}$, so $45°$ is a solution.

   $\sin\left(225°\right) + \cos\left(225°\right) = -\sqrt{2}$, so $225°$ is not a solution.

7. **Write an expression for all the solutions.**
   $$x = 45° + 360°n$$

In the next example, you need to do a little shifting at first. To solve $\cos x - \sqrt{3}\sin x = 1$, get the term with the radical in it to one side of the equation by itself. Otherwise, when you square both sides, you end up with a radical factor in one of the terms. That situation isn't always bad, but dealing with it is usually a little more awkward than not.

Solve the equation for all the possible angles from 0 to 360 degrees.

1. **Add the radical term to both sides and subtract 1 from both sides.**

   You get $\cos x - 1 = \sqrt{3}\sin x$.

2. **Square both sides.**

   $$\left(\cos x - 1\right)^2 = \left(\sqrt{3}\sin x\right)^2$$
   $$\cos^2 x - 2\cos x + 1 = 3\sin^2 x$$

3. **Replace sin² x with 1 – cos² x from the Pythagorean identity.**

   Doing so creates an equation with terms that have all the same functions, cos x, in them.

   $$\cos^2 x - 2\cos x + 1 = 3(1 - \cos^2 x)$$

4. **Simplify the equation by distributing the 3 on the right and then bringing all the terms to the left to set the equation equal to 0.**

   $$\cos^2 x - 2\cos x + 1 = 3 - 3\cos^2 x$$
   $$4\cos^2 x - 2\cos x - 2 = 0$$

5. **Divide every term by 2.**

   $$2\cos^2 x - \cos x - 1 = 0$$

6. **Factor the quadratic equation.**

   $$(2\cos x + 1)(\cos x - 1) = 0$$

7. **Set each factor equal to 0.**

   $2\cos x + 1 = 0$
   $2\cos x = -1$
   $\cos x = -\frac{1}{2}$ or
   $\cos x - 1 = 0$
   $\cos x = 1$

8. **Solve each equation for the value of x.**

   If $\cos x = -\frac{1}{2}$, then $x = \cos^{-1}\left(-\frac{1}{2}\right) = 120°, 240°$

   When you check for extraneous solutions, you find out that plugging in 120° makes the original equation false. Only 240° is a solution.

   If $\cos x = 1$, then $x = \cos^{-1}(1) = 0°, 360°$.

   The angles 0 and 360 degrees have the same terminal side. You usually list just one of them: 0 degrees.

Multiplying Through

The technique of multiplying through a trig equation by a carefully selected function shouldn't be your first choice — or your second, third, or fourth choice. This method is usually a last resort. Not that the method is terribly hard; it just requires sitting back and looking at the equation, and magically coming up with the best function to multiply through by. You can find the best function by guess or by golly, but then, that would take all the fun out of it — you want to guess right the first time. Here's an example of an equation that this technique works well on.

Solve $2 \sin x - \csc x = 1$ for all the solutions between 0 and 2π.

1. **Multiply each term by sin x.**

   Why sin x? I chose that function because I could see that the products of the individual terms would be either different powers of sine or just a number. Notice that the product of csc x and its reciprocal, sin x, is 1.

   $$2 \sin x \cdot \sin x - \csc x \cdot \sin x = 1 \cdot \sin x$$
   $$2 \sin^2 x - 1 = \sin x$$

2. **Subtract sin x from each side to set the equation equal to 0.**

   $$2 \sin^2 x - \sin x - 1 = 0$$

3. **Factor the quadratic equation.**

   $$(2 \sin x + 1)(\sin x - 1) = 0$$

4. **Set each factor equal to 0.**

   $$2 \sin x + 1 = 0 \qquad \sin x - 1 = 0$$
   $$\sin x = -\frac{1}{2} \quad \text{or} \quad \sin x = 1$$

5. **Solve for the values that satisfy the equations.**

   If $\sin x = -\frac{1}{2}$, then $x = \sin^{-1}\left(-\frac{1}{2}\right) = \frac{7\pi}{6}, \frac{11\pi}{6}$.

   If $\sin x = 1$, then $x = \sin^{-1}(1) = \frac{\pi}{2}$.

Solving with a Graphing Calculator

Some of the more-advanced graphing calculators can make short work of solving trig equations. A graphing calculator comes in very handy when the equation is complicated, has several different functions or angle multiples, or has fractional or decimal values that don't lend themselves to the traditional solving methods that I discuss throughout this chapter. For example, I prefer to use a graphing calculator to solve equations like $\cos 2x = 2\cos x$ and $\cos^2 x - 0.4 \sin x = 0.6$.

First, here's how to solve $\cos 2x = 2\cos x$ for all solutions between -2π and 2π.

1. **Put the $\cos 2x$ in the y menu (the graphing menu) of your calculator. Put the $2\cos x$ on the right as a second entry.**

   $y_1 = \cos 2x$

   $y_2 = 2\cos x$

2. **Set the window of your calculator to show the graphs.**

   Set the x values from -2π to 2π. (Be sure that your calculator is set in the radian mode.) In decimal form, let $x = -6.5$ to 6.5 to give a little room on either side of the left and right ends.

   Set the y values to go from -3 to 3. Doing so gives room above and below the graph. If you have an auto-fit capability, use it to make the graph fit automatically after you choose the x values you want the graph to encompass.

3. **Graph the two functions, and see where they intersect (see Figure 17-1).**

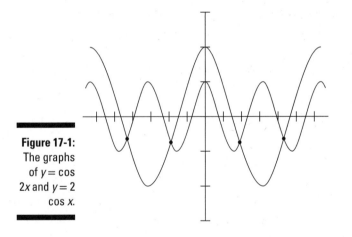

Figure 17-1: The graphs of $y = \cos 2x$ and $y = 2\cos x$.

4. **Use the intersect feature on the calculator to determine the solutions.**

The x-coordinates of the intersection points are the solutions (rounded to four decimal places): $x = -4.3377, -1.9455, 1.9455,$ and 4.3377. These solutions are in radians — the π value is already multiplied through.

You can also find the solutions to the preceding example with a graphing calculator's solver feature, but usually, you still need to look at the graph anyway so you know how many solutions you're trying to find. The solver feature usually finds only one solution at a time, and you need to give it a hint to know where to find them.

This next example has decimals built in, so you probably can't factor it. You can solve it by using identities and writing it as a quadratic, and then using the quadratic formula. This calculator method gives you another option.

Solve $\cos^2 x - 0.4 \sin x = 0.6$ for all angles between $-\pi$ and π.

1. **Put $\cos^2 x - 0.4 \sin x$ in the graphing y menu of your calculator. Put 0.6 as a second entry.**

   $y_1 = \cos^2 x - 0.4 \sin x$

   $y_2 = 0.6$

2. **Set the window of your calculator to show the graphs.**

   Set the horizontal, x values from $-\pi$ to π. In decimal form, use $x = -3.2$ to 3.2 to give a little room on either side of the ends.

   Set the vertical, y values to go from -3 to 3. Doing so gives room above and below the graph. If you have an auto-fit capability, use it to make the graph fit automatically.

3. **Graph the two functions, and see where they intersect (see Figure 17-2).**

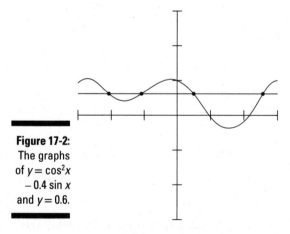

Figure 17-2:
The graphs
of $y = \cos^2 x$
$- 0.4 \sin x$
and $y = 0.6$.

4. **Use the intersect feature on the calculator to determine the solutions.**

   The x-coordinates of the intersection points are the solutions (rounded to four decimal places): $x = -2.0998, -1.0418, 0.4817,$ and 2.6598. These solutions are in radians — the π value is already multiplied through.

Chapter 18

Obeying the Laws

In This Chapter

▶ Finding missing parts in triangles

▶ Understanding the laws of sines and cosines

▶ Computing the areas of triangles

*T*riangles are very useful figures. Since humankind figured out how to keep records, people have documented the applications of triangles in mathematics and many other sciences. The right triangle gets the most use; Pythagoras saw to it that others recognized right triangles for the powerful polygons that they are. But *oblique triangles* (those that aren't right triangles) have their place, too. You can't always arrange to have a nice right triangle when you want it. Here's where oblique triangles and the laws of sines and cosines come into play.

The law of sines uses — believe it or not — the sines of a triangle's angles. With three carefully selected parts of the triangle, you can solve for the sizes of the other parts. Of course, you have to obey the law, and the choices you can make are limited. That's where the law of cosines comes in to save the day. This law isn't as user-friendly, but it picks up where the law of sines falls short.

Trigonometry opens up all sorts of possibilities for solving area problems. By using the tools in this chapter, you won't find a triangle that you can't lick.

Describing the Parts of Triangles

When it comes to triangles, you'll find right triangles and bigger-than-right triangles, called *obtuse triangles*. Other classifications exist too, such as acute, equilateral, isosceles, and so on. But no matter what you call it, any triangle has exactly six parts: three angles and three sides. After you have information about the parts of a triangle, you can perform all sorts of computations and manipulations, using triangles to model situations and solve problems.

Standardizing the parts

Usually, when you name the parts of a triangle, you follow a system or pattern. Having this system helps sort out the information, even when you don't have a picture of the triangle to help you. The most common system is to name the angles of the triangle with capital letters, usually *A*, *B*, and *C*, and name the sides opposite each of the angles with the lowercase letter that matches. Figure 18-1 shows what this labeling looks like.

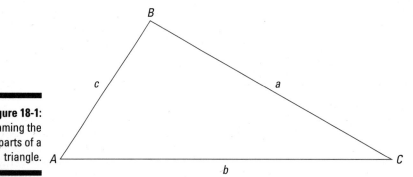

Figure 18-1: Naming the parts of a triangle.

Another common practice is to name the angles with Greek letters, such as α, β, and γ, and put those labels inside the triangle between the two sides forming the angle. But in this chapter, I stick with the capital and lowercase letters.

Determining a triangle

Even though every triangle has six parts, you only need to have information on or know the measures of three particular parts to determine the others. For example, if you know the measures of the three sides, then you know that the three angles are uniquely determined. You can't construct more than one shape and size of triangle from those three sides.

After you know the values for three carefully chosen pieces of a triangle, you can use any of the three different rules or laws that allow you to find the other three parts of the triangle. I discuss all three laws in their own sections later in this chapter.

Finding the one and only

Several combinations of parts uniquely determine a triangle. You'll probably recognize these rules from geometry, when you did proofs. Here's a list of all of the combinations you can use here.

REMEMBER

To *uniquely determine* a triangle (find only one possible shape and size), you need

- ✔ **SSS:** The measures of the three sides
- ✔ **SAS:** The measures of two sides and the angle between them
- ✔ **ASA:** The measures of two angles and the side between them
- ✔ **AAS:** The measures of two angles and one of the sides

The last rule is actually just another version of the one directly before it. When you have two angles, you can determine the third, so the side lies between two known angles. Figure 18-2 shows these situations.

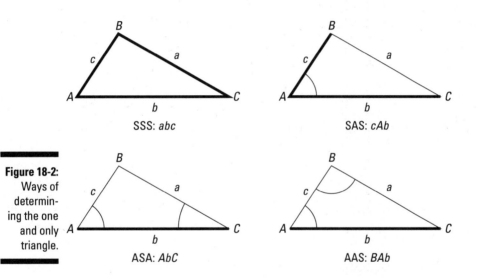

Figure 18-2: Ways of determining the one and only triangle.

SSS: *abc*

SAS: *cAb*

ASA: *AbC*

AAS: *BAb*

You may have noticed that I didn't mention one combination — AAA, where all three angles are known. I left it out on purpose because, in such a case, all you can be sure of is that the two triangles are *similar* — they're the same shape but not necessarily the same size.

Dealing with the ambiguous case

Four situations allow you to uniquely determine a triangle, and I list them in the preceding section. One other case can be helpful, even though you may end up with two different triangles instead of one unique triangle: SSA, the measures of two sides and an angle that *isn't* between them. This situation is a little tricky, because often, two different triangles are possible — which is why it's known as the *ambiguous case.* Sometimes, this case is still better than nothing — as long as you're aware that more than one triangle can exist. Figure 18-3 illustrates such a situation. In the two triangles, sides *a* and *c* and angle *A* are the same measure in each triangle. The angle measure *B* and the length of side *b*, however, aren't the same in both triangles.

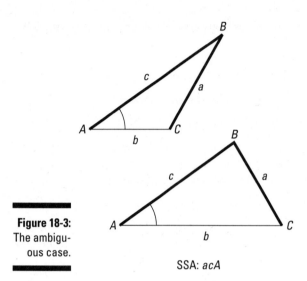

SSA: *acA*

Following the Law of Sines

When you already have two angles, as in the case of ASA or AAS (see the preceding section), you can use the *law of sines* to find the measures of the other
parts of the triangle. This law uses the ratios of the sides of a triangle and the
sines of their opposite angles. The bigger a side, the bigger its opposite angle
(and its sine). The longest side is always opposite the largest angle. Here's
how it goes.

The law of sines for triangle *ABC* with sides *a*, *b*, and *c* opposite those angles,
respectively, says

$$\frac{\sin A}{a} = \frac{\sin B}{b} = \frac{\sin C}{c} \quad \text{and} \quad \frac{a}{\sin A} = \frac{b}{\sin B} = \frac{c}{\sin C}$$

So, the law of sines says that in a single triangle, the ratio of each side to
the sine of its angle is equal to the ratio of any other side to the sine of its
angle. When working with the law of sines, you use two of the ratios at a time,
setting them equal to one another to form a proportion.

For example, consider a triangle where side *a* is 86 inches long and angles *A*
and *B* are 84 and 58 degrees, respectively. Figure 18-4 shows a picture of the
triangle, and the following steps show you how to find the missing three parts.

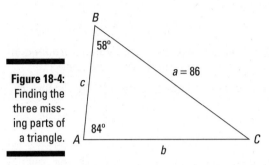

1. **Find the measure of angle C.**

   The sum of the measures of a triangle's angles is 180 degrees. So, find the sum of angles A and B, and subtract that sum from 180.

   $$180 - (84 + 58) = 180 - 142 = 38$$

   Angle C measures 38 degrees.

2. **Find the measure of side b.**

   Using the law of sines and the proportion $\dfrac{a}{\sin A} = \dfrac{b}{\sin B}$, fill in the values that you know.

   $$\frac{86}{\sin 84} = \frac{b}{\sin 58}$$

   Use the given values when writing a proportion, not those that you've determined yourself. That way, if you make an error, you can spot it easier later.

   Use the table in the Appendix or a calculator to determine the values of the sines.

   $$\frac{86}{0.995} = \frac{b}{0.848}$$

   Multiply each side by 0.848 to solve for the length b. Because the original measures are whole numbers, round this answer to the nearer whole number.

   $$(0.848)\frac{86}{0.995} = \frac{b}{0.848}(0.848)$$
   $$73.294 = b$$

   Side b measures about 73 inches.

3. **Find the measure of side c.**

   Using the law of sines and the proportion $\dfrac{a}{\sin A} = \dfrac{c}{\sin C}$, fill in the values that you know.

$$\frac{86}{\sin 84} = \frac{c}{\sin 38}$$

Again, it's best to use the given values, not those that you determined. In this case, however, you have to use a computed value, the angle C.

Use the table in the Appendix or a calculator to determine the values of the sines.

$$\frac{86}{0.995} = \frac{c}{0.616}$$

Multiply each side by 0.616 to solve for the length c. Because the original measures were given as whole numbers, round this answer to the nearer whole number.

$$(0.616)\frac{86}{0.995} = \frac{c}{0.616}(0.616)$$

$$53.242 = c$$

Side c measures about 53 inches.

Although knowing how to find the missing measures in an oblique triangle seems wonderful, you may wonder, "What's the point?" One major reason for solving triangles is so you can apply them to practical problems. For example, the question, "How tall is it?" seems to be a reasonable request.

Suppose a tree is growing on a hillside. The tree is completely vertical, but the hillside inclines at a 10-degree angle from the horizontal. Josh is standing 100 feet downhill from the tree. The angle of inclination from Josh's feet to the top of the tree is 32 degrees. How tall is the tree? First, take a look at a visual of the situation in Figure 18-5 and then review the steps that follow.

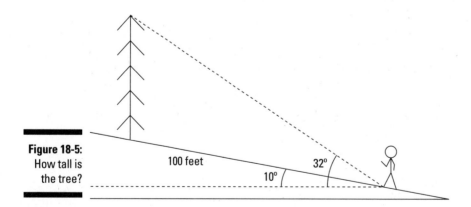

Figure 18-5:
How tall is
the tree?

100 feet

32°

10°

1. **Determine the triangle that you can use to solve the problem.**

   You know that one side is 100 feet long, and you can determine two angles, so use the triangle that Figure 18-6 shows.

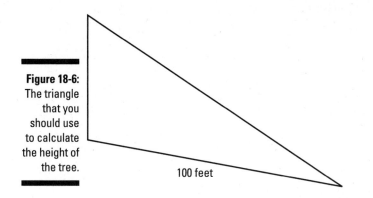

100 feet

2. **Determine the two angles on either side of the base of the triangle.**

   You determine the angle on the right of the 100-foot base by subtracting the hill's 10-degree inclination from the tree's 32-degree inclination: $32 - 10 = 22$ degrees. (These angles are also known as *angles of elevation,* which you can find out more about in Chapter 10.)

   The angle on the left of the 100-foot base is supplementary to the angle in the right triangle that you can draw below the triangle (see Figure 18-7). Drawing a right angle with the vertical leg following the tree, you determine an angle of 80 degrees by adding the 90-degree angle and the 10-degree angle and subtracting that sum from 180, the total number of degrees in a triangle. *Supplementary angles* also add up to 180, so the angle supplementary to the 80-degree angle is 100 degrees. Another way to find this 100-degree angle is to use the exterior-angle rule that follows.

   The measure of an exterior angle of a triangle is equal to the sum of the two nonadjacent interior angles.

TRIG RULES

$$1$$
$$+1$$
$$\overline{2}$$

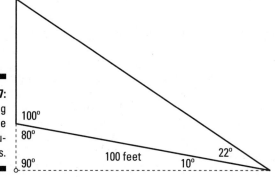

100°

80°

100 feet

22°

90°

10°

3. **Calculate the measure of the third angle.**

   Adding the two base angles together and subtracting their sum from 180 degrees, you get $180 - (22 + 100) = 180 - 122 = 58$ degrees.

4. **Determine the height of the tree.**

   The tree is the side opposite the angle measuring 22 degrees. Using the law of sines, you can write the following proportion:

   $$\frac{a}{\sin A} = \frac{b}{\sin B}$$

   $$\frac{\text{tree height}}{\sin 22} = \frac{100}{\sin 58}$$

   Solve for the height of the tree.

   $$\frac{\text{tree height}}{\sin 22} = \frac{100}{\sin 58}$$

   $$\frac{\text{tree height}}{0.375} = \frac{100}{0.848}$$

   $$(0.375)\frac{\text{tree height}}{0.375} = \frac{100}{0.848}(0.375)$$

   $$\text{tree height} = 44.222$$

   The tree is about 44 feet tall.

Continuing with the Law of Cosines

The *law of cosines* comes in handy when you have two or more sides — as in situations involving SSS and SAS — and need the measures of the other three parts. When you have two sides, you need the angle between them. If the angle isn't between the two sides, then you have the ambiguous case, SSA. Although such a situation isn't impossible, you must deal with it carefully. (See the section "Determining a triangle," earlier in this chapter, for more on these cryptic notations.)

Defining the law of cosines

The law of cosines has three different versions that you can use depending on which parts of the triangle you have measures for. Notice the pattern: The squares of the three sides appear in the equations, along with the cosine of the angle opposite one of the sides — the side set equal to the rest of the stuff.

The law of cosines for triangle *ABC* with sides *a*, *b*, and *c* opposite those angles, respectively, says

$$a^2 = b^2 + c^2 - 2bc \cos A$$

$$b^2 = a^2 + c^2 - 2ac \cos B$$

$$c^2 = a^2 + b^2 - 2ab \cos C$$

In plain English, these equations say that the square of one side is equal to the squares of the other two sides, added together, minus twice the product of those two sides times the cosine of the angle opposite the side you're solving for. Whew!

Law of cosines for SAS

When you have two sides of a triangle and the angle between them, you can use the law of cosines to solve for the other three parts. Consider the triangle *ABC* where *a* is 15, *c* is 20, and angle *B* is 124 degrees. Figure 18-8 shows what this triangle looks like.

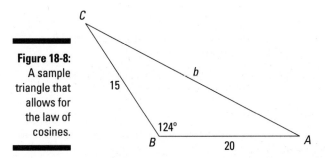

Figure 18-8: A sample triangle that allows for the law of cosines.

Now, to solve for the measure of the missing side and angles:

1. **Find the measure of the missing side by using the law of cosines.**

   Use the law that solves for side *b*.

$$b^2 = a^2 + c^2 - 2ac \cos B$$
$$= 15^2 + 20^2 - 2(15)(20)\cos 124$$
$$= 225 + 400 - 600(-0.559)$$
$$= 960.4$$

You end up with the value for b^2. Take the square root of each side and just use the positive value (because a negative length won't work here).

$b^2 = 960.4$

$b = 30.990$

The length of side b is about 31.

2. **Find the measure of one of the missing angles by using the law of cosines.**

   Using the law that solves for a, fill in the values that you know.

   $a^2 = b^2 + c^2 - 2bc \cos A$

   $15^2 = 31^2 + 20^2 - 2(31)(20)\cos A$

   Solve for $\cos A$ by simplifying, moving the other two terms to the left, and dividing by the coefficient.

   $225 = 961 + 400 - 1,240 \cos A$

   $-1,136 = -1,240 \cos A$

   $\dfrac{-1,136}{-1240} = \cos A$

   $0.916 =$

   Using the Appendix or a scientific calculator to find angle A, you find that $A = \cos^{-1}(0.916) = 23.652$, or about 24 degrees.

   You can also switch to the law of sines to solve for this angle. Don't be afraid to mix and match when solving these triangles.

3. **Find the measure of the last angle.**

   Determine angle B by adding the other two angle measures together and subtracting that sum from 180.

   $180 - (124 + 24) = 180 - 148 = 32$. Angle B measures 32 degrees.

How about an application that uses this SAS portion of the law of cosines? Consider the situation: A friend wants to build a stadium in the shape of a regular pentagon (five sides, all the same length) that measures 920 feet on each side. How far is the center of the stadium from the corners? The left part of Figure 18-9 shows a picture of the stadium and the segment you're solving for.

You can divide the pentagon into five isosceles triangles. The base of each triangle is 920 feet, and the two sides are equal, so call them both a. (Refer to the right-hand picture in Figure 18-9.) Use the law of cosines to solve for a, because you can get the angle between those two congruent sides, plus you already know the length of the side opposite that angle.

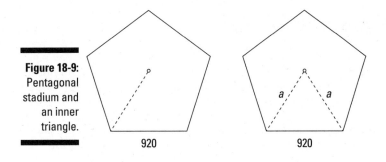

Figure 18-9:
Pentagonal
stadium and
an inner
triangle.

1. **Determine the measure of the angle at the center of the pentagon.**

   A circle has a total of 360 degrees. Divide that number by 5, and you find that the angle of each triangle at the center of the pentagon is 72 degrees.

2. **Use the law of cosines with the side measuring 920 feet being the side solved for.**

   $$c^2 = a^2 + a^2 - 2aa\cos C$$
   $$920^2 = 2a^2 - 2a^2\cos 72$$

   Because the other two sides are the same measure, write them both as *a* in the equation.

3. **Solve for the value of *a*.**

   $$920^2 = 2a^2 - 2a^2\cos 72$$
   $$846,400 = 2a^2(1 - \cos 72)$$
   $$\frac{846,400}{1 - \cos 72} = 2a^2$$
   $$\frac{846,400}{1 - 0.309} =$$
   $$\frac{846,400}{0.691} =$$
   $$1,224,891.462 =$$
   $$612,445.731 = a^2$$
   $$782.589 = a$$

   The distance from the center to a corner is between 782 and 783 feet. Now your friend knows how much fencing it'll take to divide the stadium into five equal triangles.

Law of cosines for SSS

When you know the values for two or more sides of a triangle, you can use the law of cosines. In the following case, you know all three sides but none of the angles. Solve for the measures of the three angles in triangle *ABC*, which has sides where *a* is 7, *b* is 8, and *c* is 2.

As you can see in Figure 18-10, the triangle appears to have two acute angles and one obtuse angle, the obtuse angle being opposite the longest side.

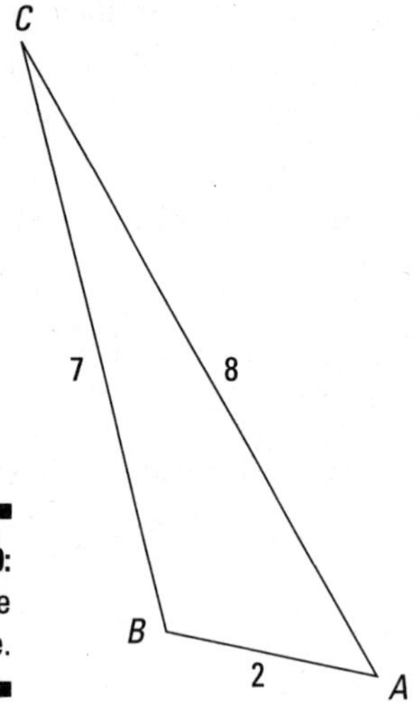

Figure 18-10:
A sample
SSS triangle.

1. **Solve for the measure of angle *A*.**

   Using the law of cosines where side *a* is on the left of the equation, substitute the values that you know and simplify the equation.

   $$a^2 = b^2 + c^2 - 2bc \cos A$$
   $$7^2 = 8^2 + 2^2 - 2(8)(2)\cos A$$
   $$49 = 64 + 4 - 32\cos A$$
   $$-19 = -32\cos A$$
   $$\frac{-19}{-32} = \cos A$$
   $$0.594 = \cos A$$

   Now use the table in the Appendix or a scientific calculator to find the measure of *A*.

$A = \cos^{-1}(0.594) = 53.559$

Angle A measures about 54 degrees.

2. **Solve for the measure of angle B.**

   Using the law of cosines where side b is on the left of the equation, input the values that you know and simplify the equation.

   $$b^2 = a^2 + c^2 - 2ac\cos B$$
   $$8^2 = 7^2 + 2^2 - 2(7)(2)\cos B$$
   $$64 = 49 + 4 - 28\cos B$$
   $$11 = -28\cos A$$
   $$\frac{11}{-28} = \cos A$$
   $$-0.393 = \cos A$$

   The negative cosine means that the angle is obtuse — its terminal side is in the second quadrant. Now use the table in the Appendix or a scientific calculator to find the measure of B.

   $A = \cos^{-1}(-0.393) = 113.141$

   Angle B measures about 113 degrees.

3. **Determine the measure of angle C.**

   Because angle A measures 54 degrees and angle B measures 113 degrees, add them together and subtract the sum from 180 to get the measure of angle C.

   $180 - (54 + 113) = 180 - 167 = 13$

   Angle C measures only 13 degrees.

Being ambiguous

Many people are visual learners, solving problems better when using a picture. This characteristic will serve such people well when it comes to solving triangles that are SSA, meaning that they know the measures of two sides and an angle that isn't between those sides. Drawing a picture helps explain why the situation may have more than one answer. When you use this setup in an actual application, the correct answer is usually pretty clear. First, I show you how to do one of these problems in general; then I show how it may actually play out in real life.

Find the missing parts of the triangle ABC that has sides a and b measuring 85 degrees and 93 degrees, respectively, and angle A measuring 61 degrees. Figure 18-11 presents the situation.

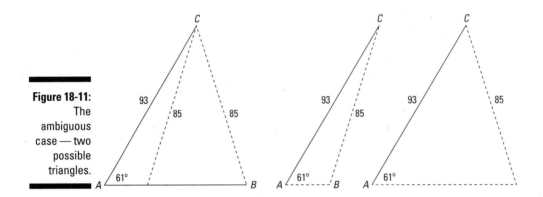

Figure 18-11: The ambiguous case — two possible triangles.

1. **Find the length of side c by using the law of cosines with a on the left-hand side of the equation.**

   Use this form because after you input the known values, it's the only one that will have just one variable to solve for — even though that variable has two powers.

   Enter the values into the law of cosines.

   $$a^2 = b^2 + c^2 - 2bc \cos A$$
   $$85^2 = 93^2 + c^2 - 2(93)(c)\cos 61$$

   Simplify the equation by performing all the operations and getting the variables alone on the right side.

   $$7,225 = 8,649 + c^2 - 186(c)(0.485)$$
   $$-1,424 = c^2 - 90.21c$$

   You end up with a quadratic equation.

   Use either the quadratic formula or a calculator to determine the solutions.

   $$0 = c^2 - 90.21c + 1,424$$
   $$c = 69.813 \text{ or } 20.397$$

   So c measures either 70 or 20.

2. **Let c measure 70, and find the measures of the other two angles.**

   This time, take a departure from the law of cosines and use, instead, the law of sines.

Use angle A and side a, and pair the ratio with angle C and side c to get the following:

$$\frac{\sin A}{a} = \frac{\sin C}{c}$$

$$\frac{\sin 61}{85} = \frac{\sin C}{70}$$

Now multiply each side by 70, and solve for the sine of C.

$$70 \cdot \frac{\sin 61}{85} = \frac{\sin C}{70} \cdot 70$$

$$70 \cdot \frac{0.875}{85} = \sin C$$

$$0.721 = \sin C$$

Solve for the angle with that sine.

$$C = \sin^{-1}(0.721) = 46.137$$

The measure of angle C is about 46 degrees.

If angle A is 61 degrees and angle C is 46 degrees, then angle B is 180 degrees minus the sum of A and C: $180 - (61 + 46) = 180 - 107 = 73$ degrees.

3. **Now let c measure 20, and find the measures of the other two angles.**

   Go back to the law of cosines to do this part. You can compare the two methods — the one in this step and the one in Step 2 — to see which one you like better.

   Use the law with c on the left-hand side of the equation to solve for the cosine of angle C:

   $$c^2 = a^2 + b^2 - 2ab\cos C$$

   $$20^2 = 85^2 + 93^2 - 2(85)(93)\cos C$$

   $$400 = 7,225 + 8,649 - 15,810\cos C$$

   $$-15474 = -15810\cos C$$

   $$\frac{-15,474}{-15,810} = \cos C$$

   $$0.979 =$$

   Use the table in the Appendix or a calculator to find the measure of angle C.

   $$C = \cos^{-1}(0.979) = 11.763$$

   Angle C measures about 12 degrees, which means that angle B is $180 - (61 + 12) = 180 - 73 = 107$ degrees.

Two wrongs make a right

When students are first introduced to fractions, they're often tempted to take some liberties with the rules that can get them into trouble. Imagine the frustration to the teacher and student alike when the student stumbles on one of the four fractions, with two digits in the numerator and denominator, where incorrect cancellation results in a correct answer. The four fractions where such a situation can occur are $\frac{64}{16}$, $\frac{98}{49}$, $\frac{95}{19}$, and $\frac{65}{26}$.

When a student mistakenly crosses out the two like digits, the result is actually the correct answer:

$$\frac{\cancel{6}4}{1\cancel{6}} = \frac{4}{1} = 4, \frac{\cancel{9}8}{4\cancel{9}} = \frac{8}{4} = 2, \frac{\cancel{9}5}{1\cancel{9}} = \frac{5}{1} = 5,$$

and $\frac{\cancel{6}5}{2\cancel{6}} = \frac{5}{2}$. Thank goodness only four such fractions exist.

The ambiguous case causes a bit of confusion. Why would you want two answers? The following example may help clear up this mystery. You really *don't* want two answers. You just want the one that answers your question.

Slim and Jim are both sitting at the intersection of two roads, which forms a 50-degree angle. They leave the intersection at the same time — Slim in his old, slow, beat-up pickup truck, and Jim in his nifty-swifty Jeep. When Jim is 400 yards down the road, the two of them are 320 yards apart. How far has Slim driven at that point?

You definitely need a picture for this problem (see Figure 18-12).

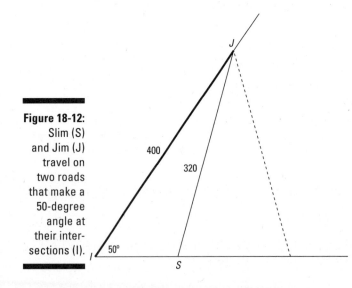

Figure 18-12: Slim (S) and Jim (J) travel on two roads that make a 50-degree angle at their intersections (I).

You can safely assume that Slim couldn't have gone farther than Jim in his old clunker — unless his truck had hidden powers. Figure out how far Slim drove, the distance from I to S (in this example, I call the distance j to be consistent with the triangle labels) by using the law of cosines. The side s is 400 yards, and angle I is 50 degrees.

1. **Write the law of cosines, and replace the letters with the values.**

$$i^2 = s^2 + j^2 - 2(s)(j)\cos I$$
$$320^2 = 400^2 + j^2 - 2(400)(j)\cos 50$$
$$102,400 = 160,000 + j^2 - 800j(0.643)$$
$$0 = 57,600 + j^2 - 514.4j$$

This equation simplifies to a quadratic equation with the variable j.

2. **Solve the quadratic equation.**

$$0 = j^2 - 514.4j + 57,600$$

Use a calculator or the quadratic formula, and you get two solutions: $x = 349.676$ and $x = 164.723$. Either answer gives you a distance smaller than the distance that Jim traveled. Refer to Figure 18-12, and choose the answer that appears to be correct.

Finding the Areas of Triangles

Finding the area of a triangle sounds relatively easy. Most grade-school children get plenty of chances to do just that. They're given a triangle and the length of the base and the height, or altitude, drawn to that base. Simple! Just plug those values into the formula, and you have it. But think about it: How many times do you have a triangular plot of land or triangular sail for a boat and have the measure of the _altitude?_

What you find in this section is a formula for every occasion. Give me a triangle, and I can find the area. Although the base and altitude would be nice, I can also do the problem with the measures of the three sides. You have two sides and an included angle? Sure, I can do that. How about two angles and an included side? I have a formula for that, too.

Of course, if you don't have the measurements for one of these exact situations, you can go to the law of sines, cosines, or tangents to fill in the blanks and find the sides or angles that you need.

Finding area with base and height

The most basic formula for finding the area of a triangle occurs when you know the base and the height. The height is drawn perpendicular to the base up to the vertex opposite that base. Figure 18-13 shows you what I mean.

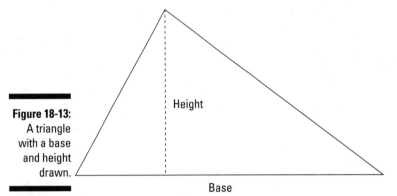

Figure 18-13:
A triangle
with a base
and height
drawn.

Height

Base

The equation for the area, *A*, of a triangle with base *b* and height *h* is $A = \frac{1}{2}bh$.

For example, to find the area of a triangle with a base measuring 12 inches and a height measuring 5 inches, input the values into the equation. You find that $A = \frac{1}{2}(12)(5) = 30$, or 30 square inches.

If the triangle happens to be a right triangle, then you're really in business. The base and height are the legs, or the two sides that are perpendicular to one another. Just find half of the base times the height. Here's an example.

Kirsten has a corner lot and wants to make a triangular garden where the two sidewalks meet. She has a 20-foot piece of border to go along the diagonal, or hypotenuse, of the triangle. She wants one side along the sidewalk to be 12 feet. How many square feet of garden will she have? Figure 18-14 illustrates the situation.

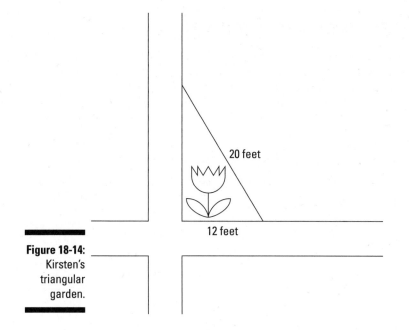

Figure 18-14:
Kirsten's
triangular
garden.

1. **Find the length of the other leg of the right triangle.**

   Using the Pythagorean theorem, and calling the missing length x, you get

   $$x^2 + 12^2 = 20^2$$
   $$x^2 + 144 = 400$$
   $$x^2 = 400 - 144 = 256$$
   $$x = \sqrt{256} = 16$$

   The other side is 16 feet long.

2. **Find the area of the triangle.**

   The base is 12 feet, and the height is 16 feet. Using the formula, you get

   $$A = \frac{1}{2}bh = \frac{1}{2}(12)(16) = 96$$

   The area is 96 square feet, which is a lot of garden to weed!

Finding area with three sides

Suppose that you have 240 yards of fencing, and you decide to build a triangular corral for your llama. Why triangular? You heard that llamas favor the shape, of course. You want the llama to have enough room to run around, so you need to know the area. What should the lengths of the triangle's sides be? You can solve this little problem by using Heron's formula for the area of a triangle.

Heron's formula says that if a triangle ABC has sides of lengths a, b, and c opposite the respective angles, and you let the semiperimeter, s, be half of the triangle's perimeter, then the area of the triangle is $A = \sqrt{s(s-a)(s-b)(s-c)}$.

In the problem of the fencing and the llama, you have many ways to make a triangular corral from 240 yards of fencing. Figure 18-15 shows a few of the possibilities. Notice that in each case, the lengths of the sides add up to 240. For the sake of this problem, don't worry about a gate.

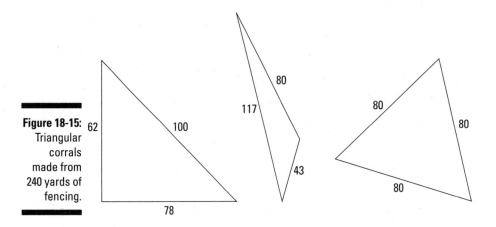

Figure 18-15: Triangular corrals made from 240 yards of fencing.

Which triangle has the greatest area? Obviously, one of them is a bit on the scrawny side, even though it uses up 240 yards of fencing, like the others. Here's how to compute the areas for the three triangles.

1. **Find the semiperimeter, s, for each triangle.**

    Referring to Figure 18-15:

      • Left triangle: $\frac{1}{2}(62+100+78) = \frac{1}{2}(240) = 120$

      • Center triangle: $\frac{1}{2}(117+80+43) = \frac{1}{2}(240) = 120$

      • Right triangle: $\frac{1}{2}(80+80+80) = \frac{1}{2}(240) = 120$

Not surprisingly, all the semiperimeters are the same, because all the perimeters are 240.

2. **Use Heron's formula to find each area.**

   Again, referring to Figure 18-15:

   - Left triangle: $A = \sqrt{120(120-62)(120-100)(120-78)}$
     $$= \sqrt{120(58)(20)(42)} = 2{,}417.933$$
   - Center triangle: $A = \sqrt{120(120-117)(120-80)(120-43)}$
     $$= \sqrt{120(3)(40)(77)} = 1{,}052.996$$
   - Right triangle: $A = \sqrt{120(120-80)(120-80)(120-80)}$
     $$= \sqrt{120(40)(40)(40)} = 2{,}771.281$$

The triangle on the right has the greatest area. Of the shapes in Figure 18-15, that triangle is the best. But you may be wondering whether another shape gives more area than that one. The answer: no. With calculus, you can prove that an equilateral triangle gives you the greatest possible area with any amount of fencing. Without calculus, you just have to try a bunch of shapes to convince yourself (or trust me).

Finding area with SAS

When you know the lengths of two of a triangle's sides plus the measure of the angle between those sides, you can find the area of the triangle. This method requires a little trigonometry — you have to find the sine of the angle involved. But the formula is really straightforward.

If triangle ABC has sides measuring a, b, and c opposite the respective angles, you can find the area with one of these formulas:

$$A = \frac{1}{2}ab\sin C \qquad A = \frac{1}{2}bc\sin A \qquad A = \frac{1}{2}ac\sin B$$

For example, look at the 30-60-90 right triangle in Figure 18-16. I use this particular example because the numbers come out so nicely.

First, find the area by using angle B and the two sides forming it.

1. **Choose the correct version of the formula.**

   The formula that uses angle B is $A = \frac{1}{2}ac\sin B$.

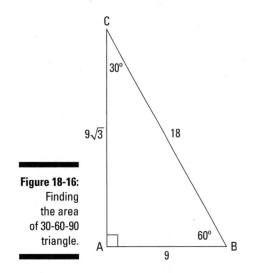

2. Find the sine of the angle.

$$\sin 60° = \frac{\sqrt{3}}{2}$$

3. Substitute the values into the formula and simplify.

$$A = \frac{1}{\cancel{2}}\left(\cancel{18}^{\,9}\right)(9)\frac{\sqrt{3}}{2} = \frac{81\sqrt{3}}{2}$$

Now find the area by using angle C and the two sides forming it.

1. Choose the correct version of the formula.

The formula that uses angle C is $A = \frac{1}{2}ab\sin C$.

2. Find the sine of the angle.

$$\sin 30° = \frac{1}{2}$$

3. Substitute the values into the formula and simplify.

$$A = \frac{1}{\cancel{2}}(\cancel{18}^{\,9})(9\sqrt{3})\frac{1}{2} = \frac{81\sqrt{3}}{2}$$

Using the method involving angle A gives you the same result, of course. For a quick comparison, just use the formula to find the area, because you're dealing with a right triangle: $A = \frac{1}{2}bh$. The methods all produce the same result.

Finding area with ASA

As you probably suspected, when you have two angles and the side between them, you can find the area of a triangle. The formulas go as follows.

In triangle ABC, if the measures of the sides are a, b, and c opposite the respective angles, you can determine the area by using one of the following equations:

$$\text{Area} = \frac{a^2 \sin B \sin C}{2 \sin A} \qquad \text{Area} = \frac{b^2 \sin A \sin C}{2 \sin B} \qquad \text{Area} = \frac{c^2 \sin A \sin B}{2 \sin C}$$

These formulas are actually built from the formula for finding the area with SAS, with a little help from the law of sines. Here's how one of them came to be.

1. **Start with the SAS rule for area.**

   $$A = \frac{1}{2}ab \sin C$$

2. **Write the law of sines involving angles A and B.**

   $$\frac{a}{\sin A} = \frac{b}{\sin B}$$

3. **Solve for b in the proportion.**

   $$\sin B \cdot \frac{a}{\sin A} = \frac{b}{\sin B} \cdot \sin B$$

   $$\frac{\sin B \cdot a}{\sin A} = b$$

4. **Substitute the equivalent for b into the area formula in Step 1.**

   $$A = \frac{1}{2}a\left(\frac{\sin B \cdot a}{\sin A}\right)\sin C$$

   $$= \frac{a^2 \sin B \sin C}{2 \sin A}$$

   That's the first formula from above. Feel free to create the others yourself.

Now consider an example. Say you have a triangle with angle A, which is 45 degrees, and angle B, which is 55 degrees, and the side between them, c, equal to 10. Find the area.

1. **Choose the correct formula — the one with c^2 in it.**

   $$A = \frac{c^2 \sin A \sin B}{2 \sin C}$$

2. **Find the sines of the two given angles.**

   The sine of 45 degrees equals 0.707, and the sine of 55 degrees equals 0.819.

3. **Find the sine of the third angle.**

   Angle C measures $180 - (45 + 55)$, or $180 - 100$, which equals 80 degrees.

   The sine of 80 degrees equals 0.985.

4. **Substitute the values into the formula and solve.**

   $$A = \frac{10^2 (0.707)(0.819)}{2(0.985)} = 29.393$$

   The area is a little over 29 square units.

Part V
The Graphs of Trig Functions

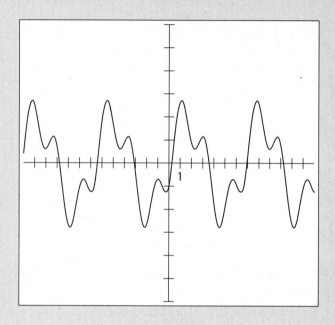

Find out more about the polar coordinate system in an article at
www.dummies.com/extras/trigonometry.

In this part...

- ✔ Create the basic graphs of the six trig functions.

- ✔ Use the basic graphs of sine and cosine to more easily graph cosecant and secant.

- ✔ Perform transformations on graphs of trig functions to make them fit a particular situation.

- ✔ Use trig functions to model periodic applications — things occurring over and over as time goes by.

Chapter 19

Graphing Sine and Cosine

In This Chapter

▶ Looking at the basic graphs of sine and cosine

▶ Working with variations of the graphs

▶ Using sine and cosine curves to make predictions

The graphs of the sine and cosine functions are very similar. If you look at them without a coordinate axis for reference, you can't tell them apart. They keep repeating the same values over and over — and the values, or outputs, are the same for the two functions. These two graphs are the most recognizable and useful for modeling real-life situations. The sine and cosine curves can represent anything tied to seasons — the weather, shopping, hunting, and daylight. The equations and graphs of the curves are helpful in describing what happens during those seasons. You also find the curves used in predator-prey scenarios and physical cycles.

The ABCs of Graphing

You can graph trig functions in a snap — well, maybe not that fast — but you can do it quickly and efficiently with just a few pointers. If you set up the axes properly and have a general knowledge of the different functions' shapes, then you're in business.

Different kinds of values represent the two axes in trig graphs. The x-axis is in angle measures, and the y-axis is in plain old numbers. The x-axis is labeled in either degrees or radians. Often, a graph represents the values from -2π to 2π to accommodate two complete cycles of the sine, cosine, secant, or cosecant functions (or four complete cycles of the tangent or cotangent functions). If the x-axis is labeled in degrees, it typically ranges from -360 to 360, which is a wide number range. That range is in sharp contrast to the y-axis, which often just goes from -5 to 5. You'll find that radians — which are real numbers — are preferable when graphing trig functions. The y-axis is labeled in real numbers;

how high and low the range extends depends on the particular function or variation of a function that you're graphing.

If you're using a graphing calculator, you need to be aware of what mode you're in when creating graphs. Otherwise, you'll get completely baffling results or none at all. For more on changing your calculator's mode, head back to Chapter 16. It just takes the press of a button or two, and you're in the right mood — oops, mode.

Waving at the Sine

The graph of the sine function is a nice, continuous wave that rolls along gently and keeps repeating itself. The domain, or x-values, of the sine function includes all angles in degrees or all real numbers in radians, so the curve has no breaks or holes. The range, or y-values, of the sine function consists of all the numbers between –1 and 1, including those two values. Figure 19-1 shows a graph of the sine function from about -2π to 2π (or from about –360 to 360 degrees).

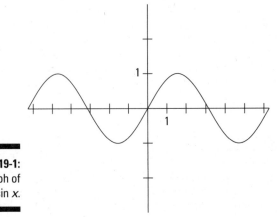

Figure 19-1:
The graph of
$y = \sin x$.

Figure 19-1 shows two complete cycles of the sine curve — the curve goes through its routine twice on the graph. If you could see the sine curve forever in either direction, it wouldn't look any different. The curve repeats the same pattern over and over again, to infinity and beyond.

Describing amplitude and period

The sine function and any of its variations have two important characteristics: the amplitude and period of the curve. You can determine these characteristics by looking at either the graph of the function or its equation.

Gaining height with the amplitude

The amplitude of the sine function is the distance from the middle value or line running through the graph up to the highest point. In other words, the amplitude is half the distance from the lowest value to the highest value. In the sine and cosine equations, the amplitude is the coefficient (multiplier) of the sine or cosine. For example, the amplitude of $y = \sin x$ is 1. To change the amplitude, multiply the sine function by a number. Take a look at Figure 19-2, which shows the graphs of $y = 3 \sin x$ and $y = \frac{1}{2} \sin x$.

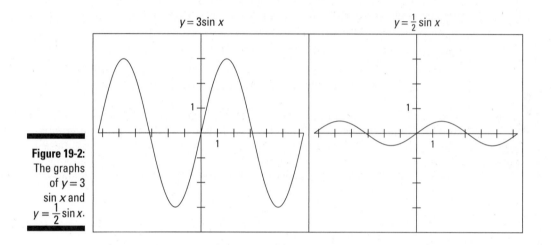

Figure 19-2: The graphs of $y = 3 \sin x$ and $y = \frac{1}{2}\sin x$.

As you can see, multiplying by a number greater than 1 makes the graph extend higher and lower. The amplitude of $y = 3 \sin x$ is 3. Conversely, multiplying by a number smaller than 1 (but bigger than 0) makes the graph shrink in value — it doesn't go up or down as far. The amplitude of $y = \frac{1}{2}\sin x$ is $\frac{1}{2}$.

The sound of music

Sounds are created by vibrations. Tuning forks can produce pure tones when they vibrate, and sine waves can model those tones. A formula for a pure tone is $y = A \sin(2\pi ft)$, where A stands for amplitude (loudness), f stands for frequency (vibrations per second), and t is a unit of time. If a string, tuning fork, or something similar vibrates at the rate of 256 times per second, then you hear middle C. When you double the frequency of any pure tone, you go up one octave, so 512 vibrations per second gives you the C above middle C.

If you add waves of different frequencies and loudness together, you get more-interesting and complex tones. The string of a violin or the inside of an oboe, for example, can vibrate with more than one frequency at the same time.

Punctuating with the period

The period of a function is the extent of input values it takes for the function to run through all the possible values and start all over again in the same place to repeat the process. In the case of the sine function, the period is 2π, or 360 degrees. Pick any place on the sine curve, follow the curve to the right or left, and 2π or 360 units from your starting point along the x-axis, the curve starts the same pattern over again.

Multiplying the angle variable, x, by a number changes the period of the sine function. If you multiply the angle variable by 3, such as in $y = \sin 3x$, then the curve will make three times as many completions in the usual amount of space. So, multiplying by 3 actually *reduces* the length of the period. In the case of $y = \sin\frac{1}{2}x$, only half the curve fits in the same space. So, a coefficient less than 1 increases the number of inputs that the function needs to complete a cycle. Figure 19-3 shows pictures of these two graphs.

The location of the multiplier makes a big difference. Multiplying the sine function by 4 and its angle variable by 4 results in two completely different graphs. The graph of $y = 4 \sin x$ is much higher than usual — the amplitude is greater than that of the standard sine function. The graph of $y = \sin 4x$ has an amplitude of 1, but the period is smaller and the curve is more scrunched together — it repeats over and over more quickly.

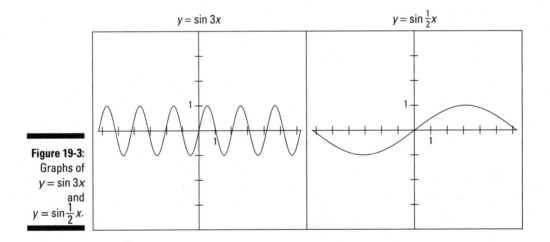

Figure 19-3:
Graphs of
$y = \sin 3x$
and
$y = \sin\frac{1}{2}x$.

Formalizing the sine equation

A general equation for the sine function is $y = A \sin B(x + C) + D$. The A and B are numbers that affect the amplitude and period of the basic sine function, respectively. The C and D create shifts in the starting and ending places and can even move the curve off the x-axis. (See the next section, "Translating the sine," for more on those movements.) When C and D are both equal to zero, you have the basic sine function $y = A \sin Bx$.

The graph of the function $y = A \sin Bx$ has an amplitude of A and a period of $\frac{2\pi}{B}$. The amplitude, A, is the distance measured from the y-value of a horizontal line drawn through the middle of the graph (or the average value) to the y-value of the highest point of the sine curve, and B is the number of times the sine curve repeats itself within 2π, or 360 degrees.

By keeping these two values in mind, you can quickly sketch the graph of this basic sine curve — or picture it in your head. For example, when graphing $y = 4\sin 2x$:

1. **Adjust for the amplitude.**

   The amplitude is 4, so the curve will extend up 4 units and down 4 units from the middle. To allow for some space above and below, set the y-axis to go from –5 to 5.

2. **Take into account the period.**

   The coefficient 2 on the x means that two complete graphs of the sine are within the space that usually houses only one.

3. **Graph the curve from -2π to 2π (see Figure 19-4).**

   You can see that the graph goes from –4 to 4 and that four complete cycles are in the space that usually houses only two.

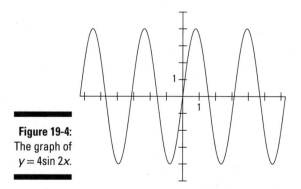

Figure 19-4:
The graph of
$y = 4\sin 2x$.

Translating the sine

Playing around with the amplitude and period of the sine curve can result in some interesting changes to the basic curve. That curve is still recognizable, though. You can see the rolling, smooth curve crossing back and forth over a middle line. In addition to those changes, you have two other options for altering the sine curve — shifting the curve up, down, or sideways. These shifts are called *translations* of the curve. (Turn back to Chapter 3 for a basic discussion on translating functions.) And the translations are accounted for in the more general equation for the sine: $y = A \sin B(x + C) + D$.

Sliding up or down

You can move a sine curve up or down by simply adding or subtracting a number from the equation of the curve. In terms of the equation, if D is positive, you move the curve upward that amount; if D is negative, it goes down. For example, the graph of $y = \sin x + 4$ moves the whole curve up 4 units, with the sine curve crossing back and forth over the line $y = 4$. On the other hand, the graph of $y = \sin x - 1$ slides everything down 1 unit. Figure 19-5 shows what the two graphs look like.

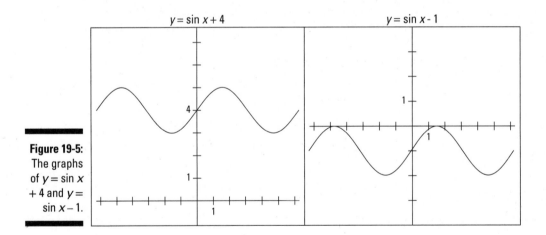

Figure 19-5:
The graphs
of $y = \sin x$
$+ 4$ and $y =$
$\sin x - 1$.

As you can see, the basic shape of the sine curve is still recognizable — the curves are just shifted up or down on the coordinate plane.

Shifting left or right

By adding or subtracting a number from the angle in a sine equation, you can move the curve to the left or right of its usual position. This is the C part of the general equation. This shift, or translation, relates the sine curve to the cosine curve. But the translation of the sine itself is important: Shifting the curve left or right can change the places that the curve crosses the x-axis or some other horizontal line. For example, the graph of $y = \sin (x + 1)$ is the usual sine curve slid 1 unit to the left, and the graph of $y = \sin(x - 3)$ slid 3 units to the right. Figure 19-6 shows the graphs of the original sine equation and these two shifted equations.

Take a look at the point marked on each graph in Figure 19-6. This point illustrates how an *intercept* (where the curve crosses an axis) shifts on the graph when you add or subtract a number from the angle variable.

WARNING!

Note the difference between adding or subtracting a number to the function and adding or subtracting a number to the angle measure. These operations affect the curve differently, as you can see by comparing Figure 19-5 and Figure 19-6.

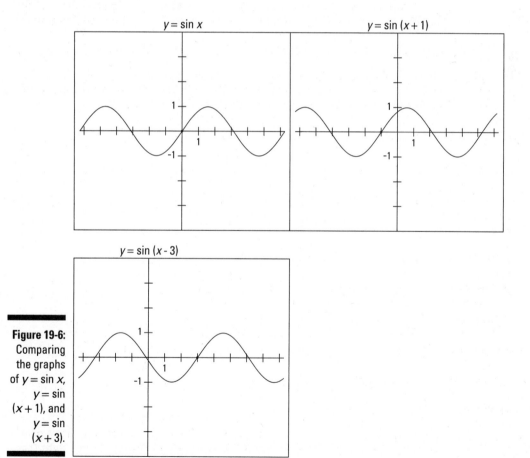

Figure 19-6: Comparing the graphs of $y = \sin x$, $y = \sin (x + 1)$, and $y = \sin (x + 3)$.

$y = \sin x + 2$ Adding 2 to the function raises the curve by 2 units.

$y = \sin (x + 2)$ Adding 2 to the angle variable shifts the curve 2 units to the left.

Graphing Cosine

The graph of the cosine function looks very much like that of the sine function. This quality is due to the fact that they're related by domain and range, as well as by several identities. An identity involving a shift explains the relationship best, because that shift can make the graph of the sine function look like the cosine function.

Comparing cosine to sine

The relationship between the sine and cosine graphs is that the cosine is the same as the sine shifted to the left by 90 degrees, or $\frac{\pi}{2}$. The equation that represents this relationship is $\cos x = \sin\left(x + \frac{\pi}{2}\right)$. Look at the graphs of the sine and cosine functions on the same coordinate axes, as shown in Figure 19-7. Each tick mark on the x-axis represents one unit. The graph of the cosine is the darker curve; note how it's shifted to the left of the sine curve.

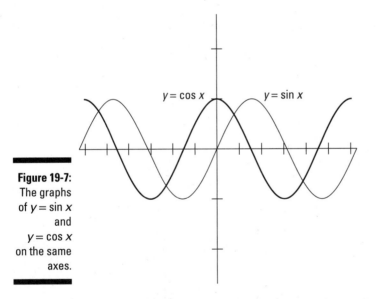

Figure 19-7:
The graphs
of $y = \sin x$
and
$y = \cos x$
on the same
axes.

The graphs of the sine and cosine functions illustrate a property that exists for several pairings of the functions. This property is based on the right triangle and the two acute or complementary angles in a right triangle. The identities that arise from the triangle are called the *co-function identities*.

TRIG RULES
1
+1
2

The co-function identities are

$$\sin \theta = \cos(90° - \theta) \qquad\qquad \csc \theta = \sec(90° - \theta)$$
$$\cos \theta = \sin(90° - \theta) \qquad\qquad \sec \theta = \csc(90° - \theta)$$
$$\tan \theta = \cot(90° - \theta) \qquad\qquad \cot \theta = \tan(90° - \theta)$$

These identities show how the function values of the complementary angles in a right triangle are related. For example, $\cos \theta = \sin(90° - \theta)$ means that if θ is equal to 25 degrees, then $\cos 25° = \sin(90° - 25°) = \sin 65°$. This equation is a roundabout way of explaining why the graphs of sine and cosine are different by just a slide. You probably noticed that these co-function identities all use the difference of angles, but the slide of the sine function to the left was a sum. The shifted sine graph and the cosine graph are really equivalent — they become graphs of the same set of points.

Using properties to graph cosine

The cosine function has the same amplitude and period as the sine function: The amplitude is 1, and the period is 2π, or 360 degrees. The variations on the cosine work the same way as on the sine. If you want to change the amplitude, multiply the cosine function by a number. If you want to change the period, multiply or divide the angle variable by a number. To slide the whole curve up, down, right, or left, add or subtract a number from the whole function or the angle variable.

For the equation $y = 3\cos x$, the amplitude is 3, meaning the graph stretches up and down to 3 units. The graph of $y = \cos (x - 3)$ is shifted 3 units to the right. Figure 19-8 shows you the graphs.

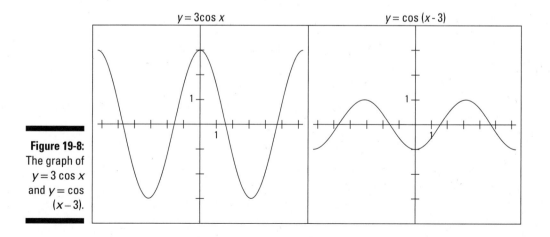

$y = 3\cos x$ $y = \cos (x - 3)$

Figure 19-8:
The graph of
$y = 3 \cos x$
and $y = \cos$
$(x - 3)$.

The graphs of sine and cosine are difficult to tell apart when they're shifted about. But that fact just shows how much those functions have in common, which can work to your advantage when you're applying them.

Applying the Sines of the Times

The sine curve and its co-function, cosine, are great for modeling situations that happen over and over again in a predictable fashion. Some examples include the weather, seasonal sales of goods, body temperature, the tide's height in a harbor, average temperatures, and so on. In this section, I show you a few examples of how you can use these functions in practical situations. In each case, I point out how the graph and formula illustrate the amplitude, period, and any shifts (for more on those concepts, check out the "Waving at the Sine" section, earlier in this chapter).

Sunning yourself

San Diego, California, is a gorgeous part of the world. Whether it's summer or winter, you want to be there. But what if you're someone who likes long, sunny days? When is the best time to go there? Assume that the following formula gives you the number of hours of daylight in San Diego when you input any day of the year. Letting t be the day of the year (from 1 to 365), you can figure the number of hours of sunlight, H, with the equation $H(t) = 2.4 \sin(0.017t - 1.377) + 12$. Figure 19-9 shows the graph of this equation.

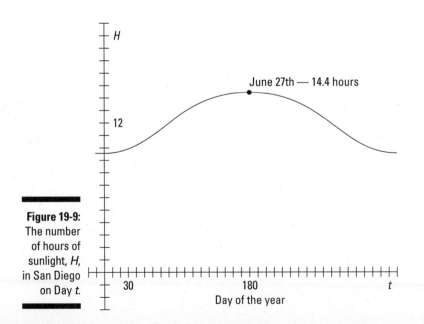

Figure 19-9:
The number of hours of sunlight, H, in San Diego on Day t.

Trusty old protractors

A *protractor* is a familiar instrument to grade-school and high-school students. They use this flat, semicircular instrument, which is marked with degrees from 0 to 180 degrees, to construct and measure angles.

The protractor has been around for a very long time. The first protractors were used in navigation to plot the positions of ships on navigational charts. In 1801, Joseph Huddart, a U.S. Navy captain, invented an instrument called a *three-arm protractor* or *station pointer.*

The amplitude of the sine curve is 2.4, which means that the number of daylight hours extends 2.4 hours above and below the average number of daylight hours. The average number of daylight hours is 12, which is the translation upward. When you add or subtract the 2.4, you find that the hours of sunlight range from 14.4 to 9.6, depending on the time of year. The period is $\frac{2\pi}{0.017} \approx 370$, which is a little longer than a year because of the rounding in the formula. The coefficient (multiplier) on the t in the function $H(t) = 2.4\sin(0.017t - 1.377) + 12$ means that 0.017 of the curve takes up the usual amount of space for one curve, 2π units. As you can see on the graph, the day with the most sunlight is June 27. You can determine that high point by using a graphing calculator that finds it for you — as well as the y-value of when t equals 14.4, or you can use calculus to solve it! Do you know, now, when you want to go to San Diego?

Averaging temperature

A relatively reasonable model for the average daily temperature in Peoria, Illinois, is $T(x) = 50 - 42\cos(0.017x - 0.534)$, where x is the day of the year starting with January 1 as Day 1. The $T(x)$ represents the temperatures in degrees Fahrenheit. The graph is in radian measure, because radians are real numbers, as opposed to degrees — you need numbers to count off the days. Figure 19-10 shows what the graph of the function looks like for the whole year.

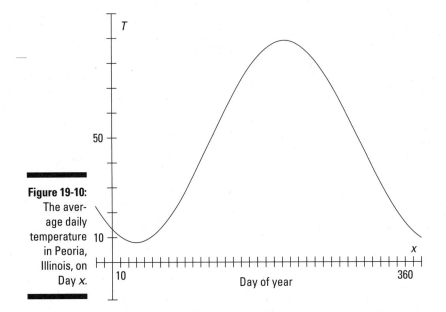

The multiplier on the cosine function is 42, so the amplitude of the curve is
42. Don't worry about the negative sign in front of the 42. The curve goes
upward and downward anyway, so the negative sign just makes it go downward and then upward, instead of the reverse.

The period is affected by the multiplier 0.017. The result of that multiplication is that only 0.017 of a cosine curve takes up the usual amount of space
for an entire curve, which is 2π, or a little over 6 units. Because this graph is
for a whole year, the curve has to spread out over 365 units, so that each of
the horizontal units has just a little part of it.

The shift upward of 50 units is the middle or average temperature for the
year. Add the amplitude of 42 to this number, and the average temperature
gets up to 92 degrees; subtract the amplitude, and the average gets down to
8 degrees. Note that the curve starts a little to the right of the *y*-axis to
account for when the seasons change. If you want more details on curve
translations to the left, right, up, and down, go to Chapter 22.

What do you do with the graph? You can estimate when the highest and lowest
temperatures occur and get an idea of the types of temperatures to expect if you
move to Peoria, Illinois. Figure 19-11 shows the graph of the average temperatures with points for some days of the year and the average temperatures on
those days. A graphing calculator is indispensable when graphing these figures
and calculating values. You can either input *x*-values to find out what the *y*-values
are at those points, or you can trace along the curve to get the measures.

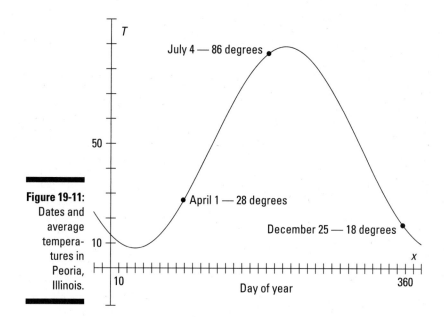

Figure 19-11: Dates and average temperatures in Peoria, Illinois.

July 4 — 86 degrees

April 1 — 28 degrees

December 25 — 18 degrees

50

10

10

360

Day of year

Taking your temperature

The temperature of a person's body fluctuates during the day instead of staying at a normal 98.6 degrees. And actually, not everyone has a "normal" temperature. Lots of people run either hot or cold.

If you're one of the special people with a normal temperature, then your temperature goes up and down by about 1 degree each day. The formula $T(x) = \sin(x + 0.262) + 98.6$ may be a model of your temperature during a 24-hour period. The variable x is the number of hours since midnight, so this equation uses a 24-hour clock. The temperatures are given in degrees Fahrenheit. The graph is in radians, so you can enter the numbers for the hours. Figure 19-12 shows what a graph of the temperatures may look like, noting a few times and temperatures. Now you see why your feet get cold in the wee hours of the morning.

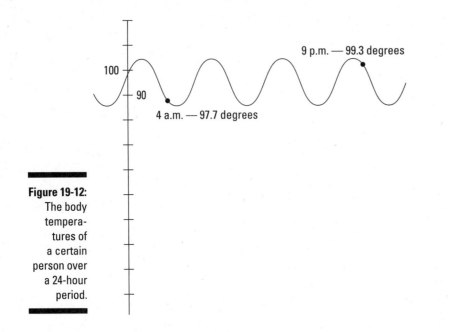

9 p.m. — 99.3 degrees

4 a.m. — 97.7 degrees

Figure 19-12:
The body
tempera-
tures of
a certain
person over
a 24-hour
period.

Making a goal

Even though people in many parts of the world play soccer year-round, certain times of the year show an increase in the sales of outdoor soccer shoes. Here's a model for the sales of pairs of shoes where N is in millions of pairs and m is the month of the year: $N(m) = 44 \sin(0.524m) + 70$. From the equation, you can tell that the average number of pairs sold is 70 million, which is the vertical shift upward. That number fluctuates between 26 million and 114 million, which you find by adding and subtracting the amplitude, 44, to and from the average. The period of this model is $\dfrac{2\pi}{0.524} \approx 11.99$, or 12 months. Figure 19-13 shows a graph of this function.

A graph like the one in Figure 19-13 can help distributors and retailers with their plans for sales.

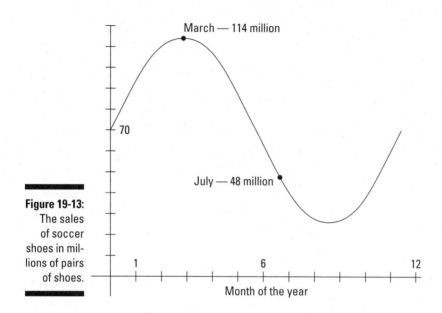

March — 114 million

70

July — 48 million

1   6   12

Month of the year

Figure 19-13:
The sales
of soccer
shoes in mil-
lions of pairs
of shoes.

Theorizing with biorhythms

Many years ago, the public showed great interest in a person's *biorhythms*,
which are the physical, emotional, and intellectual cycles that a person experiences in life. Many people even wrote books about them. Some believe that
these cycles affect how a person reacts to situations in his or her life. They
even go so far as to say that the positions of these curves have influenced
major decisions of famous movie stars and politicians.

What do biorhythms have to do with trigonometry? Everything! This biorhythm theory uses the sine curve. Supposedly, our life cycles start at birth
and fluctuate like sine curves. The physical cycle is 23 days long, the emotional cycle is 28 days long, and the intellectual cycle is 33 days long. If you
plot all these cycles on a graph, starting on the day you were born, you can
see where these cycles are right now and what they'll look like in the future.
Figure 19-14 shows a graph of the three biorhythm cycles starting on the day
a person is born.

In Figure 19-14, you can see how the different cycles have different periods.
Imagine these sine curves going on for years and years, crossing over the *x*-
axis and over one another. Figure 19-15 shows some biorhythm cycles plotted
for some imaginary person for some year in the month of March.

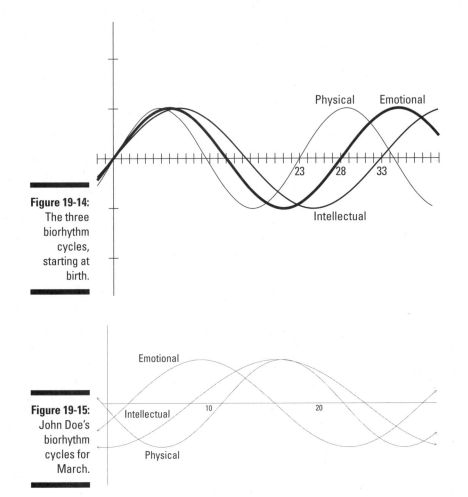

Figure 19-14: The three biorhythm cycles, starting at birth.

Figure 19-15: John Doe's biorhythm cycles for March.

If you believe in the biorhythm theory that says these curves exist, you can see that on about the 13th of the month, all the cycles are above the *x*-axis, and after about the 25th, they're all below the *x*-axis. Supposedly, when a curve is above the *x*-axis, everything is bright and sunny — a person is in good health, emotionally fine, and very smart and with it. When the curve is below the *x*-axis, the person tends to be sick, depressed, and dull. In addition, the theory says that when the cycles cross from above to below the axis, or vice versa, those days are critical. A critical day is when upheaval and crises are possible. Such a day is a good time to stay in bed — if that's even safe. I guess there's no way to prove or disprove this theory, but it sure makes interesting use of the sine curve!

Chapter 20

Graphing Tangent and Cotangent

. . . .

In This Chapter

▶ Comparing tangent and cotangent

▶ Indicating lines where one curve ends and another begins

▶ Moving a graph up, down, and all around

. . . .

*T*he tangent and cotangent functions have lots of similarities. You can write both functions in terms of sine and cosine, so they share the same function values in their ratios. One difference between tangent and cotangent is that they don't have function values in the same places for the *x*-values in their domain — they shift over by 90 degrees. Even though their domains (or *x*-values) are restricted, tangent and cotangent are the only trig functions with ranges (or *y*-values) that go all the way from negative infinity to positive infinity. The challenges in graphing tangent and cotangent are in dealing with the domain restrictions and *asymptotes* (dotted vertical lines used to determine the shape of a curve), as you see in this chapter.

Checking Out Tangent

The tangent function can be written as the ratio of the sine divided by the cosine: $\tan\theta = \frac{\sin\theta}{\cos\theta}$. (For more information on the tangent function, see Chapters 7 and 8.) The sine and cosine functions have values for every *x*-value, so no matter what number you put in for *x*, you'll get an answer. The only problem occurs when the cosine function is equal to 0, because a fraction can't have a 0 in the denominator. So, wherever the cosine function is equal to 0, the graph of the tangent curve doesn't exist, and this is indicated with asymptotes. Another interesting property comes into play with the fact that the sine and cosine are both positive in the first quadrant and negative in the third quadrant. As a result, the tangent is positive in those two quadrants and negative in the other two, because either the sine or cosine is negative, but not both.

Determining the period

The sine and cosine functions have a period of 2π, or 360 degrees, which means that after every 2π, the function pattern starts all over again. In the case of the tangent function, though, the length of the period is only π — half as long as that of sine or cosine. The tangent function repeats its pattern over and over twice as frequently as sine and cosine.

Assigning the asymptotes

An *asymptote* is a line that helps give direction to a graph. This line isn't part of the function's graph; instead, it helps determine the shape of the curve by having the curve hug or get very close to the asymptote. Asymptotes are usually indicated with dashed lines. If you use your graphing calculator, though, to graph a tangent curve, the asymptotes will appear as solid lines. The calculator seemingly wants to keep everything connected. Just remember that the asymptote is just there for form.

The asymptotes for the graph of the tangent function occur regularly, each of them π, or 180 degrees, apart. They separate each piece of the tangent curve, or each complete cycle from the next.

The equations of the tangent's asymptotes are all of the form $x = \frac{\pi}{2}(2n+1)$, where n is an integer. Under that stipulation for n, the expression $2n + 1$ always results in an odd number. By replacing n with various integers, you get lines such as $x = \frac{\pi}{2}$, $x = \frac{3\pi}{2}$, $x = \frac{5\pi}{2}$, $x = \frac{7\pi}{2}$, $x = -\frac{\pi}{2}$, $x = -\frac{3\pi}{2}$, $x = -\frac{5\pi}{2}$, and $x = -\frac{7\pi}{2}$. The reason that asymptotes always occur at these odd multiples of $\frac{\pi}{2}$ is because those points are where the cosine function is equal to 0. As such, the domain of the tangent function includes all real numbers *except* the numbers that occur at these asymptotes.

Figure 20-1 shows what the tangent function looks like when graphed. The tangent values go infinitely high as the angle measure approaches 90 degrees, 270 degrees, and so on (as you move from left to right on the graph). The values go infinitely low as the angle measure approaches –90 degrees, –270 degrees, and so on (as you move from right to left on the graph).

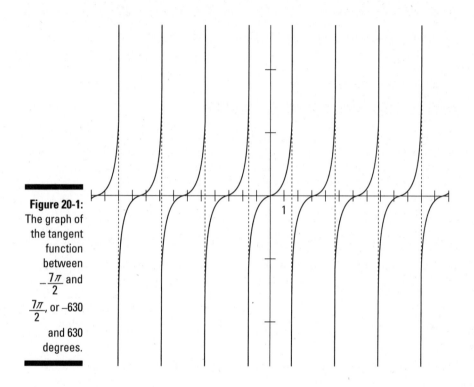

Figure 20-1:
The graph of
the tangent
function
between
$-\dfrac{7\pi}{2}$ and
$\dfrac{7\pi}{2}$, or –630
and 630
degrees.

As you can see, the tangent function repeats its values over and over. One main difference between this function and the sine and cosine functions is that the tangent has all these breaks between the cycles. As you move from left to right, the tangent appears to go up to positive infinity. It actually disappears at the top of the graph and then picks up again at the bottom, where the values come from negative infinity. Graphing calculators and other graphing utilities don't usually show the graph disappearing at the top, so it's up to you to know what's actually happening, even though the picture may not look exactly that way.

Because graphing calculators try to connect the tangent function to make it continuous across the screen, you get a false impression of any curve with vertical asymptotes. The only way to get rid of those extra lines is to turn your calculator to the *dot* mode (as opposed to the *connected* mode). Most calculators have ways to set the settings (or mode) for things such as degrees and radians, dotted graphs and connected graphs, floating decimals and fixed decimals, and so on. The changes are usually easy to make — just see your calculator's manual for specific instructions. The hard part is remembering what setting you're in.

Fiddling with the tangent

You can alter the tangent function with multiplication, addition, and subtraction. In some cases, the effects are similar to those that occur when you alter the sine and cosine functions. Because these results aren't similar all the time, you should consider the alterations on a case-by-case basis.

Multiplying the tangent

You can multiply the tangent function by a number, but doing so doesn't affect the function the way that it affects the sine function. Multiplying the sine by a number changes its amplitude, making the function include larger and smaller values. The tangent values, however, already go from negative infinity to positive infinity.

When you multiply the entire tangent function by a number, here's what happens:

- ✔ **If you multiply by a number bigger than 1,** the graph of the function gets steeper more quickly.

- ✔ **If you multiply by a fraction between 0 and 1,** the graph of the function gets flatter.

- ✔ **If you multiply by a negative number,** the curve flips over the *x*-axis. For more on these flips (called *reflections*), go to Chapter 3.

Figure 20-2 shows graphs of the basic tangent function ($y = \tan x$) and two multiples to illustrate this property. Notice how the multiplier of 6 makes the tangent curve steeper, whereas the multiple of 0.2 makes it flatten out. Both functions still have values that go from negative infinity to positive infinity, but the rate at which they get there changes.

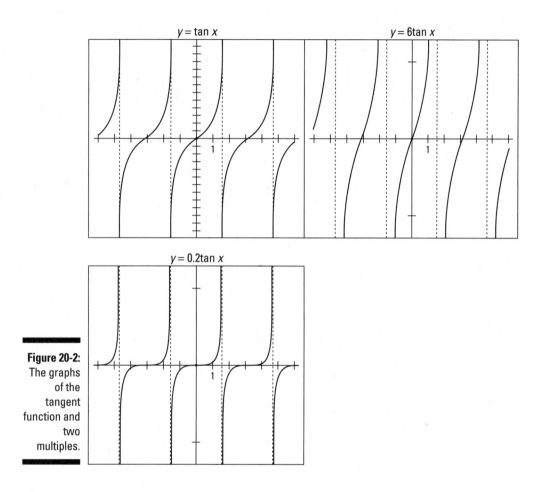

Figure 20-2:
The graphs
of the
tangent
function and
two
multiples.

Multiplying the angle

Multiplying the angle variable in the tangent function has the same effect as it does with the sine and cosine functions. If the multiple is 2, as in $y = \tan 2x$, then the tangent function makes twice as many cycles in the usual amount of space. In other words, the period is $\frac{\pi}{2}$, which is the tangent's usual period, π, divided by 2. Because multiplying the angle variable of the tangent function mirrors the results of doing the same with the sine and cosine functions, I don't go into detail here — for more information, refer to Chapter 19.

Figure 20-3 shows a few graphs to illustrate the effect of multiplying the angle variable by a number greater than 1 and then by a number between 0 and 1.

The graph of $y = \tan 3x$ doesn't show all the asymptotes, but that graph has three times as many tangent curves as usual. The graph of $y = \tan\frac{1}{2}x$ has only half as many cycles — or it takes twice as long to complete one cycle.

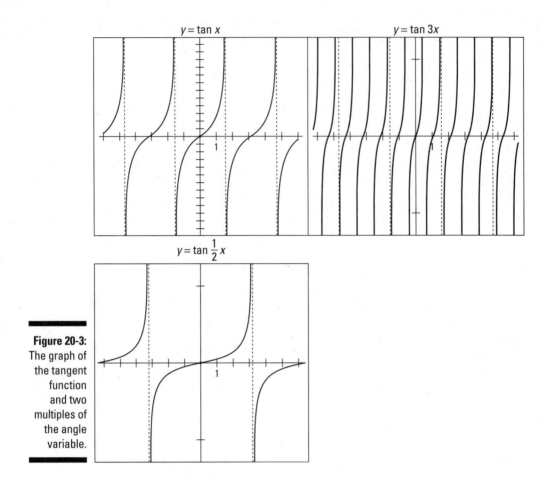

Adding to tangent

Adding a number to the tangent function results in raising the curve on the graph by that amount. Likewise, subtracting a number drops the curve. Because the tangent function has values from negative infinity to positive infinity, adding to or subtracting from the function doesn't change what values the tangent has — it just changes where they happen. When you add or subtract, the *point of inflection* in the tangent curve (where the curve appears to flatten out a bit) shifts up or down. Figure 20-4 shows some graphs to illustrate this shift.

Adding or subtracting a number from the angle variable of the tangent function has the same effect as with the sine and cosine — it moves the curve to the left or right. The graph of $y = \tan (x + 1)$ shifts one unit to the left, including the asymptotes. The graph of $y = \tan (x - 1)$ moves everything to the right one unit. Figure 20-5 shows a comparison of the tangent function and the two shifted curves.

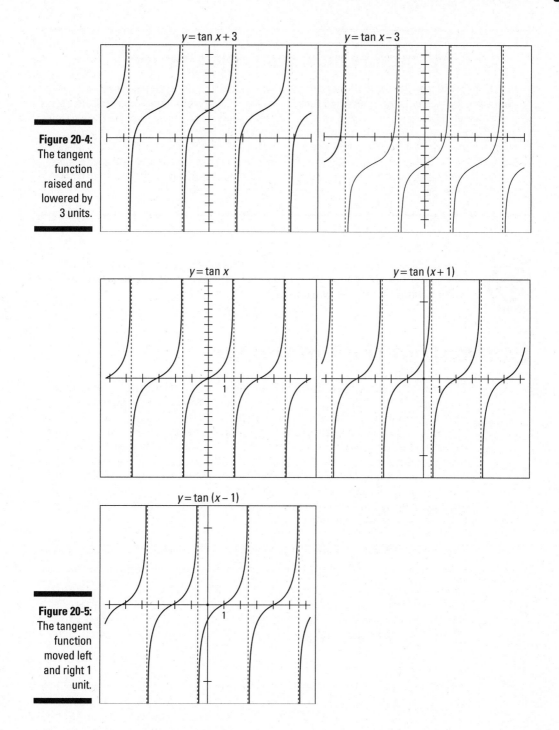

Figure 20-4:
The tangent function raised and lowered by 3 units.

$y = \tan x + 3$

$y = \tan x - 3$

$y = \tan x$

$y = \tan (x + 1)$

$y = \tan (x - 1)$

Figure 20-5:
The tangent function moved left and right 1 unit.

Wherefore didst thou come, radian?

In 1873, a man named James Thomson defined and named the *radian,* the angle measure equivalent to about 57 degrees. Thomson was a mathematics professor at Queens College in Belfast, Northern Ireland. He was the brother of the famous physicist William Thomson, also known as Lord Kelvin. Although James's work seems to affect more people directly — everyone who studies or uses radian measure — his brother gained more recognition. William, Lord Kelvin, was also a mathematician who used mathematics to connect physics and electrostatics. He was the target of T. H. Huxley, an evolutionist who had some issues with mathematics and claimed that Kelvin underestimated the age of Earth.

 If you have a tough time telling these graphs apart, just look for the point of inflection of the tangent curve. The point of inflection is a good reference mark when looking at all these variations.

Confronting the Cotangent

The graphs of the tangent function lay the groundwork for the graphs of the cotangent. After all, they're cofunctions and reciprocals, and have all sorts of connections. The two graphs are similar in so many ways: They both have asymptotes crossing the graph at regular intervals, go from negative infinity to positive infinity in value, and are affected by multiplying and adding. The biggest difference is in the direction the graphs are drawn. The values of the tangent function appear to *rise* as you read from left to right. The function goes upward, disappears off the graph, and then reappears down below to start all over again. The cotangent function does the opposite — it appears to *fall* when you read from left to right.

The asymptotes of the cotangent curve occur where the sine function equals 0, because $\cot\theta = \frac{\cos\theta}{\sin\theta}$. Equations of the asymptotes are of the form $x = n\pi$, where n is an integer. Some examples of the asymptotes are $x = -3\pi$, $x = -2\pi$, $x = -\pi$, $x = 0$, $x = \pi$, $x = 2\pi$, and $x = 3\pi$. (For an explanation of asymptotes, refer to the section "Assigning the asymptotes," earlier in this chapter.) Figure 20-6 shows the cotangent function graphed between -3π and 3π.

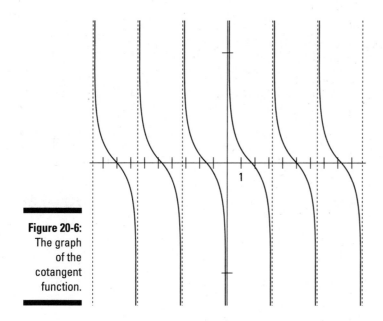

Figure 20-6:
The graph
of the
cotangent
function.

Like the other functions, cotangent repeats the same values over and over. You can apply the same types of variations to cotangent that you can to tangent (refer to the section "Fiddling with the tangent" for the details). Figure 20-7 depicts three examples of variations: multiplying the angle variable, subtracting from the function, and adding to the angle variable.

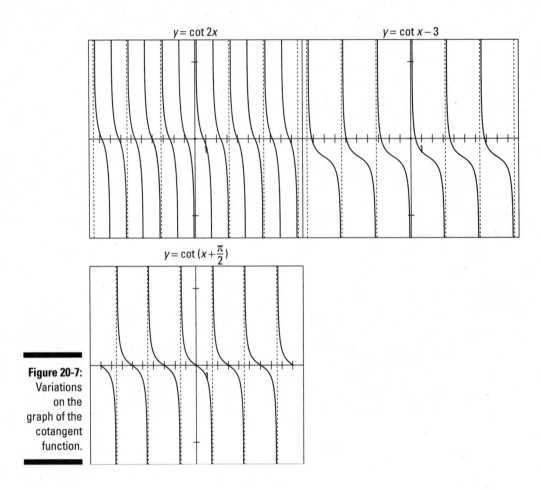

$y = \cot 2x$

$y = \cot x - 3$

$y = \cot \left(x + \dfrac{\pi}{2}\right)$

Figure 20-7:
Variations
on the
graph of the
cotangent
function.

Chapter 21

Graphing Other Trig Functions

- -

In This Chapter

▶ Using sine and cosine to graph their reciprocals

▶ Drawing the inverse functions on a graph

- -

*T*he functions cosecant and secant have similarities to one another not only because they're the reciprocals of sine and cosine, but also because their graphs look very much alike. As you see in this chapter, the easiest way to sketch the graphs of these two functions is to relate them to the graphs of their reciprocals. Doing so helps determine the *asymptotes* (where the curve approaches infinity or negative infinity), turning points, and general shape of the curves.

Seeing the Cosecant for What It Is

The cosecant function is the reciprocal of the sine function (meaning, the cosecant equals 1 divided by the sine). Even though the sine function has a domain that includes every possible number, that characteristic can't be true of its reciprocal. Whenever the sine function is equal to 0, the cosecant function doesn't exist. That fact helps determine the asymptotes you use to graph the cosecant function.

Identifying the asymptotes

The domain of the cosecant function is any number except multiples of π, because those measures are where the sine function is equal to 0. You can use this situation to identify the asymptotes by simply writing equations that use multiples of π. The asymptotes of the cosecant function are of the form $x = n\pi$, where n is some integer. Some examples of the asymptote equations are $x = -3\pi$, $x = -2\pi$, $x = -\pi$, $x = 0$, $x = \pi$, $x = 2\pi$, $x = 3\pi$, and $x = 4\pi$.

Using the sine graph

One really efficient way of graphing the cosecant function is to first make a quick sketch of the sine function. With that sketch in place, you can draw the asymptotes through the *x-intercepts* (where the curve crosses the *x*-axis). These are the places where sin *x* = 0. You can also use the maximum and minimum values on the sine function to locate the minimum and maximum points (known as *turning points*) of the cosecant function.

To graph *y* = csc *x*:

1. **Sketch the graph of *y* = sin *x* from –4π to 4π, as shown in Figure 21-1.**

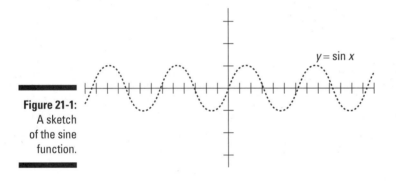

Figure 21-1:
A sketch of the sine function.

2. **Draw the vertical asymptotes through the *x*- intercepts, as Figure 21-2 shows.**

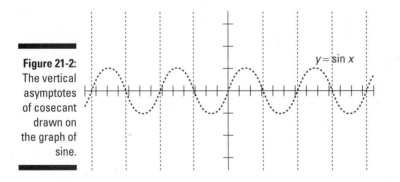

Figure 21-2:
The vertical asymptotes of cosecant drawn on the graph of sine.

3. **Draw $y = \csc x$ between the asymptotes and down to (and up to) the sine curve, as shown in Figure 21-3.**

   The cosecant goes down to the top of the sine curve and up to the bottom of the sine curve. The sine and cosecant share those points where the y-values are 1 and –1.

   After using the asymptotes and reciprocal as guides to sketch the cosecant curve, you can erase those extra lines, leaving just $y = \csc x$. Figure 21-4 shows what this function looks like all on its own.

The range of the cosecant function includes all values equal to or greater than 1 and all values equal to or less than –1. In Figure 21-4, you can see that a gap in function values lies between 1 and –1. The cosecant curve, just like all the other trig functions, keeps repeating its pattern over and over.

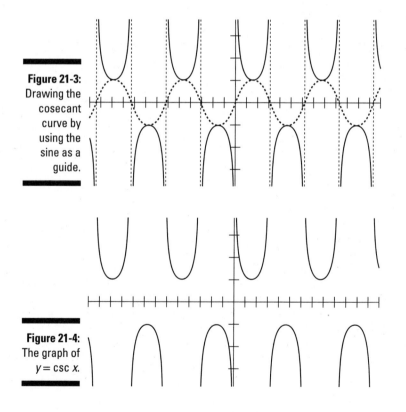

Figure 21-3: Drawing the cosecant curve by using the sine as a guide.

Figure 21-4: The graph of $y = \csc x$.

Varying the cosecant

How can you make changes to the cosecant function? This function is affected by the same multiplication, addition, and subtraction principles that affect the other functions (check out Chapter 19 for more-detailed info).

Adding or subtracting a number to or from the *cosecant function* results in slides of the graph up or down. Adding or subtracting numbers to the *angle variable* slides the graph left or right. And now I get right to it and do two slides for the price of one, sliding the graph to the left by 2 units and up by 2 units. The equation of that graph is $y = \csc (x + 2) + 2$. To find out why adding to the angle, x, moves the graph left, head on back to Chapter 3. Meanwhile, Figure 21-5 shows the graph of this equation.

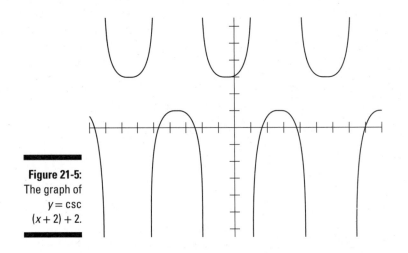

Figure 21-5:
The graph of
$y = \csc$
$(x + 2) + 2$.

Although I left out the asymptotes, you can still tell where they are — the shape of the graph is pretty clear.

Multiplying by a number changes the steepness and period of the cosecant function. If you multiply the function by 2, the curve gets steeper and has more space between its bottom and top. If you multiply the angle variable by 2, twice as much of the curve fits in the usual amount of horizontal space. Figure 21-6 shows both changes in the graph of $y = 2 \csc 2x$.

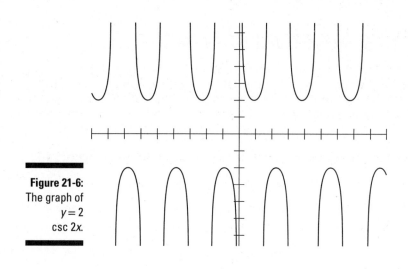

Figure 21-6:
The graph of
$y = 2$
csc $2x$.

Unveiling the Secant

The techniques that you use to graph the secant curve parallel those that you use to graph the cosecant. First, identify the asymptotes by determining where the reciprocal of secant — cosine — is equal to 0. Then sketch in that reciprocal, and you can determine the turning points and general shape of the secant graph.

Determining the asymptotes

Because the secant equals 1 divided by the cosine, the secant function is *undefined,* or doesn't exist, whenever the cosine function is equal to 0. You can write the equations of the asymptotes by setting y equal to those values where the cosine is equal to 0, so the asymptotes are $x = -\dfrac{7\pi}{2}$, $x = -\dfrac{5\pi}{2}$,

$x = -\dfrac{3\pi}{2}$, $x = -\dfrac{\pi}{2}$, $x = \dfrac{\pi}{2}$, $x = \dfrac{3\pi}{2}$, $x = \dfrac{5\pi}{2}$, $x = \dfrac{7\pi}{2}$, and so on. Another way

to express the equations of all the asymptotes is to write $x = \dfrac{(2n+1)\pi}{2}$, where

n is some integer.

Sketching the graph of secant

Using the graph of the cosine to sketch the graph of the secant function is the easiest method. Graph the cosine very lightly or with a dotted curve — the same as with the asymptotes. A lot of busywork is associated with this graph, but you just have to ignore all the extra stuff and zoom in on the graph that you want. To sketch the graph of the secant function:

1. **Sketch the graph of $y = \cos x$ from -4π to 4π, as shown in Figure 21-7.**

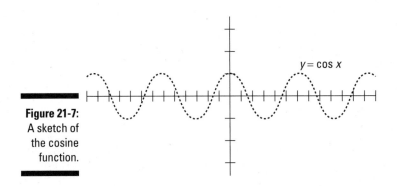

Figure 21-7:
A sketch of the cosine function.

2. **Draw the vertical asymptotes through the x-intercepts (where the curve crosses the x- axis), as Figure 21-8 shows.**

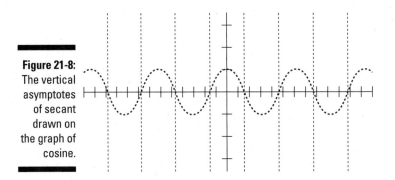

Figure 21-8:
The vertical asymptotes of secant drawn on the graph of cosine.

3. **Draw $y = \sec x$ between the asymptotes and down to (and up to) the cosine curve, as shown in Figure 21-9.**

The secant goes down to the top of the cosine curve and up to the bottom of the cosine curve — where the cosine has a value of 1 and –1, respectively.

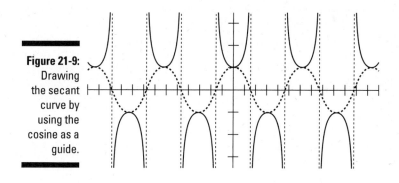

Figure 21-9:
Drawing
the secant
curve by
using the
cosine as a
guide.

Fooling around with secant

The secant graph is different from the cosecant in several ways, but one of the most obvious ways is that this graph is symmetric about the y-axis. The secant is a mirror reflection over that axis. You can use this property to do something interesting to the graph.

The usual translations and multiplications (refer to Chapter 3) affect the secant graph. If you multiply the function by $\frac{1}{6}$ and add 2π to the angle variable, as in the equation $y = \frac{1}{6}\sec(x + 2\pi)$, Figure 21-10 shows what happens.

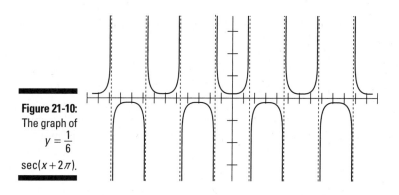

Figure 21-10:
The graph of
$$y = \frac{1}{6}$$
$\sec(x + 2\pi)$.

Compared to $y = \sec x$, the graph in Figure 21-10 is much closer to the x-axis and seems to be flattened out between the asymptotes. These changes happen when you multiply the function by a number between 0 and 1. The turning point is still in the same place, but the y-value is much closer to 0.

Now you see it; now you don't

One optical illusion, called the Kanizsa Triangle, causes the eye to perceive a white equilateral triangle where none is actually drawn. Here it is:

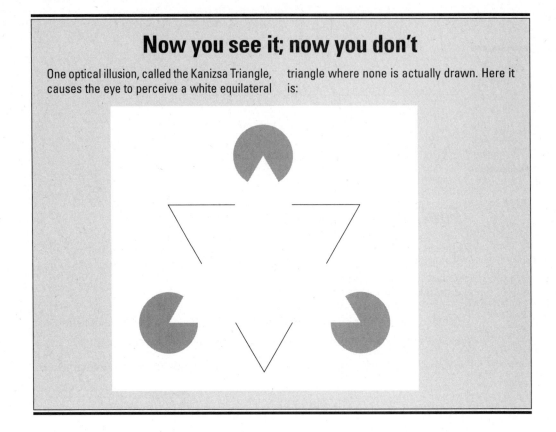

The other curiosity is that the asymptotes don't seem to be different. They aren't — and they shouldn't be. By adding 2π to the angle variable, you shift the graph 2π units to the left. The graph really has shifted, but you can't tell, because the new graph lies completely on the old one. When the shift is equal to the *period* of the function (the length of the interval that it takes for the function values to start repeating over again), the change isn't apparent.

Laying Out the Inverse Functions

The six basic trig functions all have inverses. In Chapter 15, you find information on the notation used to indicate inverse functions, what their respective domains are, and how to use them.

The inverse trig functions — $y = \sin^{-1}x$, $y = \cos^{-1}x$, $y = \tan^{-1}x$, $y = \cot^{-1}x$, $y = \sec^{-1}x$, and $y = \csc^{-1}x$ — are useful when solving trigonometric equations or doing applications involving trigonometry. The graphs of the inverse trig functions are rather unique; inverse sine and inverse cosine are rather abrupt and disjointed, but inverse tangent and inverse cotangent seem to go on forever, within narrow confines. The reason you find these big differences is because of the range or outputs of the original functions. The range of sine and cosine is between –1 and 1, so the inverse function will have inputs of just those values. The tangent and cotangent have infinite ranges — which is why their inverses have infinite domains.

Why in the world are the graphs of inverse functions of any importance? For the same reason that all pictures are important — for their visual impact. Especially in the world of trig functions, remembering the general shape of a function's graph goes a long way toward helping you remember more about the function values and using them effectively.

Before diving into this section, you may want to go back and review the material on inverse functions in Chapter 3 if you need to reacquaint yourself with the domains and ranges of these functions and their respective values.

Graphing inverse sine and cosine

The first two graphs sort of go together — they have a common characteristic. The input values for both $y = \sin^{-1}x$ and $y = \cos^{-1}x$ are all the numbers from –1 to 1, including those numbers. The inputs are restricted to those values because they're the *output* values of the sine and cosine.

The output, or range, values for these two inverse functions are different. The range of $y = \sin^{-1}x$ consists of angles in the first and fourth quadrants. In radians, the range is $-\frac{\pi}{2}$ to $\frac{\pi}{2}$; in approximate decimal values, the range is –1.571 to 1.571. The range of $y = \cos^{-1}x$, on the other hand, consists of angles in the first and second quadrants, or angles from 0 to π. In approximate decimal values, that range is 0 to 3.142.

Figure 21-11 shows what the graphs of inverse sine and cosine look like.

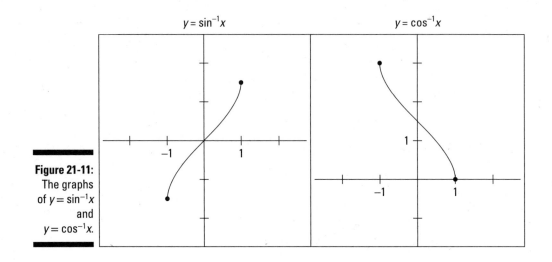

$y = \sin^{-1}x$

$y = \cos^{-1}x$

The points indicated on the graphs are at $x = -1$ and $x = 1$. These points are the extreme values of the inputs. The y-values represent the angle measures. If you want to find a point on either graph, just find some number between -1 and 1, and find the place on the graph corresponding to that x-value.

Taking on inverse tangent and cotangent

The tangent and cotangent functions have restricted inputs — certain angles don't jibe with them. But their outputs go through all the real numbers. If you switch those two groups of numbers to fit the inverses of tangent and cotangent, you can say that the *inputs* go through all the real numbers, and the *outputs* are restricted. The graphs of these two inverse functions are quite interesting because they both involve two horizontal asymptotes. The asymptotes help with the shapes of the curves and emphasize the fact that some angles won't work with the functions.

The two horizontal asymptotes for the inverse tangent function are $y = -\frac{\pi}{2}$ and $y = \frac{\pi}{2}$, because the tangent function doesn't exist for those two angle measures. The tangent function isn't defined wherever the cosine is equal to 0. If you need to review the tangent function, go to Chapters 7 and 8. The graph of the inverse tangent has x-values from negative infinity to positive infinity, with all y-values between those two asymptotes.

The two horizontal asymptotes for the inverse cotangent function are $y = 0$ and $y = \pi$. As with the inverse tangent, the inverse cotangent function goes from negative infinity to positive infinity between the asymptotes. Check out both graphs in Figure 21-12.

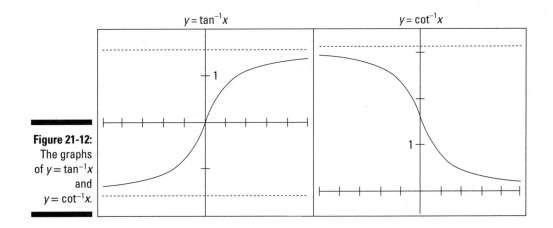

$y = \tan^{-1}x$ $\qquad\qquad$ $y = \cot^{-1}x$

Figure 21-12:
The graphs
of $y = \tan^{-1}x$
and
$y = \cot^{-1}x$.

The main differences between these two graphs is that the inverse tangent curve *rises* as you go from left to right, and the inverse cotangent curve *falls* as you go from left to right. Also, the horizontal asymptotes for inverse tangent capture the angle measures for the first and fourth quadrants; the horizontal asymptotes for inverse cotangent capture the first and second quadrants. The measures between these asymptotes are, of course, consistent with the ranges of the two inverse functions.

Crafting inverse secant and cosecant

The graphs of the inverse secant and inverse cosecant will take a little explaining. First, keep in mind that the secant and cosecant functions don't have any output values (*y*-values) between –1 and 1, so a wide-open space plops itself in the middle of their graphs. This idea translates into a wide-open space between the *x*-values –1 and 1 in the graphs of their inverses. Also, the graphs of secant and cosecant go infinitely high and infinitely low along the *y*-axis. So, the graphs of the inverses have a horizontal asymptote. All this talk probably seems like nonsense, so take a look at Figure 21-13, which shows the graphs.

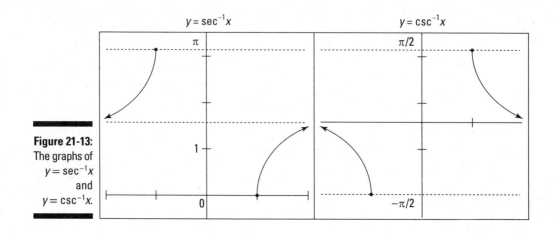

Figure 21-13:
The graphs of
$y = \sec^{-1}x$
and
$y = \csc^{-1}x$.

The graph of $y = \sec^{-1}x$ lies between 0 and π on the y-axis. All the output values are in the first and second quadrants. But a horizontal asymptote runs through the graph: the line $y = \frac{\pi}{2}$. The secant isn't defined at $\frac{\pi}{2}$, so its inverse won't have an output value there. The graph of the inverse secant goes from the point $(1,0)$ and moves upward, staying below the horizontal asymptote as the x-values go to positive infinity. It also comes from negative infinity along the x-axis above the horizontal asymptote, moving upward to the point $(-1, \pi)$.

The graph of $y = \csc^{-1}x$ lies between $-\frac{\pi}{2}$ and $\frac{\pi}{2}$, with a horizontal asymptote of $y = 0$. (The cosecant isn't defined at $x = 0$, so its inverse doesn't have an output value there.) The graph of inverse cosecant covers angle measures from the first and fourth quadrants. On the right, the graph goes from the point $\left(1, \frac{\pi}{2}\right)$ down toward the horizontal asymptote as the x-values go to positive infinity. On the left, the graph's x-values come from negative infinity, where they're just below the asymptote, and move down to the point $\left(-1, -\frac{\pi}{2}\right)$.

Chapter 22

Topping Off Trig Graphs

The graphs of the trigonometric functions can take on many variations in their shapes and sizes. As wonderful as these graphs are just by themselves, they're even better and more useful when you adjust them to fit a particular situation. In Chapters 19, 20, and 21, I show you how to make the trig functions slide about by moving them up, down, left, and right. I also show you how to make them steeper and flatter. In this chapter, I complete the trig story with additional transformations, as well as the even-more-exciting possibilities that occur when you combine graphs. I start off with a basic template for a trig function and progress from that point.

The Basics of Trig Equations

You can identify all the different transformations that you can perform on a trig function from a certain form of the function's equation. First, check out the general equation and then consider some examples of what the specific equations may look like.

TRIG RULES
1 +1 2

The general form for a trig equation is $y = Af\left[B(x+C)\right]+D$, where

▶ f represents the trig function.

▶ A represents the *amplitude,* or steepness.

• A positive A means the graph is oriented as usual.

• A negative A means that the graph is flipped over a horizontal line.

▶ B determines the *period* of the graph (the length of the interval needed for the graph of the function to start repeating itself) using the formula $\frac{2\pi}{B}$ or $\frac{\pi}{B}$, depending on the function's usual period.

✔ *C* determines a shift to the left or right.

✔ *D* determines a shift up or down.

Here are some examples of trig functions using this format:

$$y = -2\sin\left[4\left(x + \frac{\pi}{4}\right)\right] + 3$$

$$y = \frac{1}{2}\cos\left[\frac{1}{6}(x - \pi)\right] + 1$$

$$y = -\cot 4x + \frac{1}{2}$$

Each of the numbers changes the basic graph in a particular way.

In the graph of $y = -2\sin\left[4\left(x + \frac{\pi}{4}\right)\right] + 3$, you have a graph that has four complete cycles of the sine curve in the space where you would usually find one. The graph has a highest value of 5 (adding the 2 to the 3), is shifted to the left slightly, and goes downward where the sine curve usually goes upward. Does this seem a bit mysterious? If so, refer back to Chapter 19 for more details on sleuthing out the particulars.

Ready for another? In the graph of $y = \frac{1}{2}\cos\left[\frac{1}{6}(x - \pi)\right] + 1$, the graph seems a bit stunted, if you just look between -2π and 2π. That's because there's only $\frac{1}{6}$ of the usual cosine curve in the space that you usually find the whole curve.

This curve is also only half the usual height, and it's slid up by 1 unit and to the right by about 3. It would be very difficult to try to create the function equation if you had just the graph to go by.

The last one is quick and easy. The graph of $y = -\cot 4x + \frac{1}{2}$ has four complete cotangent cycles where you'd usually find one. It's also "flipped" over a horizontal axis, and it's raised by $\frac{1}{2}$ unit.

In the following sections, I give you some tools to work with in figuring out the graphs of these types of trig functions.

Flipping over a horizontal line

When you multiply the trig function by a negative number, all the output values are reversed. The positive values become negative, and the negative values become positive. The effect that this operation has on the graph is that it appears to have a reflection or flip over a horizontal line. For example, Figure 22-1 shows the graph of $y = -\sin x$ compared to $y = \sin x$.

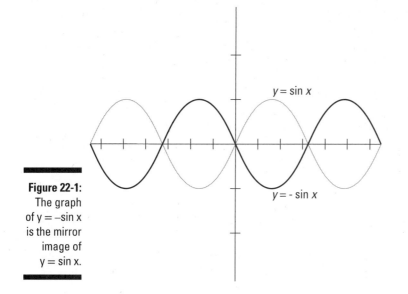

Figure 22-1:
The graph
of y = −sin x
is the mirror
image of
y = sin x.

$y = \sin x$

$y = -\sin x$

See how the two graphs compare? The original graph appears to be flipped over the *x*-axis.

Interpreting the equation

Each of the different letters of the general equation for a trig function has a purpose. Here's a more-detailed explanation of each part.

A is for amplitude

The letter *A* represents the amplitude of the sine or cosine function, and it affects the steepness or flatness of the graphs of any of the trig functions. If the absolute value (ignore the + or – sign) of *A* is some number greater than 1, then the graph is steeper than usual. If the absolute value of *A* is between 0 and 1, then the graph is flatter. The higher the number, the steeper the curve. The closer the number is to 0, the flatter the curve.

B is for becoming (the period)

The multiplier *B* affects the length of the graph's *period,* or how far it goes along the *x*-axis. The sine, cosine, cosecant, and secant all normally have a period of 2π. The tangent and cotangent have a period of π. If you divide the normal period of the function by the value of *B*, you get the length of the new, adjusted period. Another way to put it: *B* tells you how many complete cycles the curve will make in the space that usually has only one. If *B* is 2, the graph has two complete cycles where there's usually one.

C is for cruisin' left or right

The value of C changes the graph by moving the whole curve to the left or right of where it usually is. If you subtract C, the graph moves C units to the right. If you add C, it moves C units to the left.

D is for distancing yourself up or down

The value of D tells how far up or down the graph moves from its original position. A positive D moves the graph up, and a negative D moves it down. The value of D also represents the average or middle value of the sine and cosine curves and the middle of the open space of the secant and cosecant curves.

Graphing with the General Form

Now is the time to put all your knowledge to work and do some serious graphing. In the general equation for a trig function, $y = Af\left[B(x+C)\right]+D$, the letters A, B, C, and D all represent values, but they have to be in those exact places, and the equation has to be in that exact form. You need to factor, multiply, or manipulate the equation in other ways to get it in the general form if you want to use these values to figure out what the graph looks like.

More likely than not, you'd draw the graphs in this section with the help of a graphing calculator. The only problem with graphing calculators is that entering these complicated functions correctly is often a real challenge. You can't use brackets or braces to help keep the groupings straight. You're stuck with parentheses, which can get messy when you're dealing with a lot of them. The main reason I provide the examples in this section is so you know what to expect. When you know how all these variations work, you're able to recognize when you have an error in your graphing calculator work and avoid that age-old saying, "Garbage in, garbage out."

The first example involves graphing $y = 3\sin\left[2\left(x-\frac{\pi}{4}\right)\right]+1$.

1. **Determine the amplitude of the curve.**

   The 3 represents the A, which is the amplitude of a sine curve. The function will stretch 3 units above and below the middle. The 3 is positive, so the curve doesn't flip or reflect over a horizontal line.

2. **Find the period of the function.**

The 2 represents the B, which means that the curve makes two complete cycles in the amount of space where it usually has only one. Because the normal period of the sine function is 2π units, in this function the period is $\frac{2\pi}{2}$, or π units long.

3. Determine the shift left or right.

The $-\frac{\pi}{4}$ represents the value of C. Because C is negative, the shift is $\frac{\pi}{4}$ units to the right.

4. Find the shift up or down.

The last number, 1, is the value of D, which is a shift upward by 1 unit.

5. Input all the values to graph the equation, as shown in Figure 22-2.

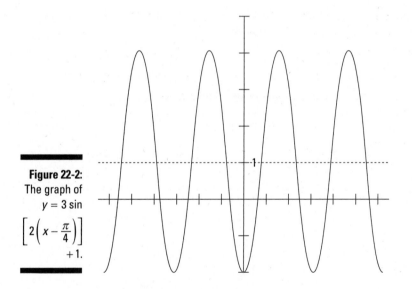

Figure 22-2: The graph of $y = 3\sin\left[2\left(x - \frac{\pi}{4}\right)\right] + 1.$

In Figure 22-2, I drew the line $y = 1$ to show the middle, so you can see the result of the vertical shift. The graph shown goes from about -2π to 2π on the x axis, where you'd normally expect to find two complete cycles. Instead, the graph has four.

The next graph has a flip over a horizontal line. The curve doesn't flip over the x-axis because the graph is dropped down by 3 units. Instead, the flip is over the horizontal line $y = -3$. Without further ado, here's how to sketch the graph of $y = -2\cos\left[\frac{1}{3}(x + \pi)\right] - 3.$

1. **Determine the amplitude.**

   The –2 in front tells you two things. First, the amplitude is 2, or the curve is twice as high as usual. The negative sign tells you that the whole cosine curve is flipped over a horizontal line. Where the curve usually goes up, it goes down, and vice versa.

2. **Find the period.**

   The multiplier of $\frac{1}{3}$ spreads the curve out quite a bit — only one-third as much curve is in the same amount of space as a 2π period (the period of the basic cosine function) usually has. In fact, the new period is

   $$2\pi \div \frac{1}{3} = 2\pi \cdot 3 = 6\pi.$$

3. **Determine the shift left or right.**

   The value of C is π, which is a little more than 3 units. The graph moves 3 units to the left, because C is positive.

4. **Find the vertical shift.**

   D represents the number –3, so the whole graph shifts down 3 units.

5. **Now use all these values to graph the curve from -2π to 2π, as Figure 22-3 shows.**

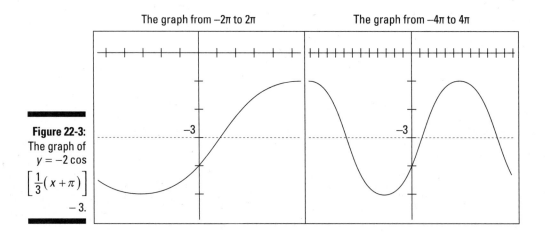

The graph from –2π to 2π   The graph from –4π to 4π

Figure 22-3:
The graph of
$y = -2 \cos$
$\left[\frac{1}{3}(x + \pi)\right]$
$- 3.$

As you can see from the graphs in Figure 22-3, just graphing from -2π to 2π doesn't show a complete cycle. The period is 6π, so the graph needs more space. The two graphs look a little different — the one on the right looks steeper, because the scales on the x-axes are different. Both graphs are of the same function. The dotted line is $y = -3$, which is the middle or average of the graph.

The seed of life

A *compass* is an instrument that you can use to draw a circle. All you do is place the sharp, pointed end at the center of the circle and drag the pencil end around. One of the basic constructions that students can produce is a set of circles that intersect with one another in an interesting pattern. This pattern is formed by marking six equidistant points on the original circle, using the radius set on the compass from that circle. Then six new circles are formed using the six points as centers. When constructed correctly, those circles form what's called the *seed of life,* an arrangement that has ancient Egyptian beginnings. This pattern is also a part of 13th-century Italian art. Here's what it looks like:

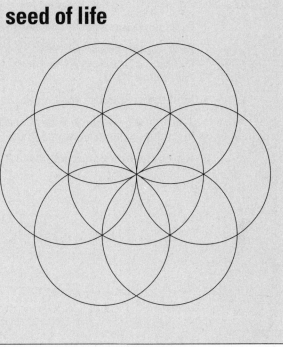

This last graph shows that you don't always have to have a value for one of the letters in the general form. Well, actually, each of the letters always has a value, but when that value is 1 or 0, it doesn't show up. Just know that the part of the equation with the value of 1 or 0 doesn't change the basic graph. In the graph of $y = -\tan 2x + \pi$, you find two situations where the original graph doesn't change.

1. **Determine the steepness.**

   With the tangent function, I don't refer to the multiplier as *amplitude* because the tangent curve doesn't have a highest or lowest point, as the sine and cosine curves do. Any multiplier *A* affects the steepness. In this case, that steepness doesn't change, because the *A* is essentially a 1. Because the 1 is negative, the graph flips over a horizontal line.

2. **Find the period of the function.**

   The multiplier of 2 on the angle measure makes the period of this tangent curve equal to $\frac{\pi}{2}$, because the normal period of the tangent function is π, and you have to divide by 2 here. The graph makes twice as many cycles in the usual amount of space.

3. **Determine the shift left or right.**

Here's another case where the graph doesn't change. The equation has no number in place of C — that value is actually 0. So, the graph doesn't shift left or right.

4. **Find the vertical shift.**

The number π is the D value. That number is positive, so the graph shifts up π units, which is about 3 units.

5. **Graph the function from -2π to 2π, as Figure 22-4 shows.**

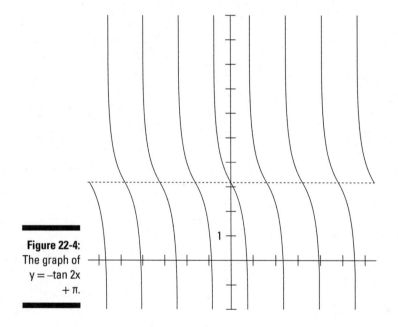

Figure 22-4:
The graph of $y = -\tan 2x + \pi$.

You can see that the graph in Figure 22-4 shows eight complete cycles of the tangent function. The reason it shows so many cycles is because the graph goes from -2π to 2π, and each cycle is only $\frac{\pi}{2}$ in length. The dotted line shows the horizontal shift of π; I left the vertical asymptotes out.

Adding and Subtracting Functions

Just when you thought this book couldn't get any better, I add yet another twist to the trigonometry picture. You can model many applications in physics and the cycles of nature with curves that you create by adding or subtracting two trig functions together or by adding a trig function and some algebraic function. When you add functions together, you can obtain the graph of this sum by taking each x-value from each function and finding the sum of the y-values that correspond to those x-values. Then you can plot the points with the x-values that you used and the y-values that you found. I show you a couple of examples, and because they get too messy very quickly, I bail and use a graphing calculator.

The function $y = x + \sin x$ is the sum of the sine function $y = \sin x$ and the algebraic function $y = x$. The algebraic function $y = x$ is a line that cuts diagonally through the third and first quadrants. The sine has y-values that go from -1 to 1, over and over again. Table 22-1 shows some of the separate functions' values and then the sum of those values.

Table 22-1			The y-Values of $y = x + \sin x$								
x	-2π	$-\pi$	-2	-1	$-\frac{\pi}{2}$	0	$\frac{\pi}{2}$	1	2	π	2π
sin x	0	0	-0.909	-0.841	-1	0	1	0.841	0.909	0	0
x + sin x	-2π	$-\pi$	-2.909	-1.841	-2.571	0	2.571	1.841	2.909	π	2π

You can get a better idea of how this addition works by looking at the graph. I used a graphing calculator to graph $y = x + \sin x$ from -2π to 2π. And then, to give you an even better picture of what's going on, and because I think this curve is neat, I graphed it from -4π to 4π in Figure 22-5.

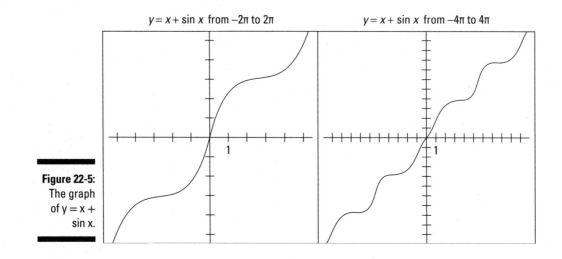

$y = x + \sin x$ from -2π to 2π $y = x + \sin x$ from -4π to 4π

Figure 22-5:
The graph
of y = x +
sin x.

The next example shows what can happen when you subtract one trig function from another. Of course, I had to experiment with all sorts of different combinations of functions to make this graph come out especially interesting. You should try your hand at it, too. Here's my contribution, the graph of $y = 2\sin x - \cos 3x$. I subtracted the function values of each, one from the other, to produce the very nice curve in Figure 22-6, which I drew from -4π to 4π. Is it a heartbeat or a pretty design?

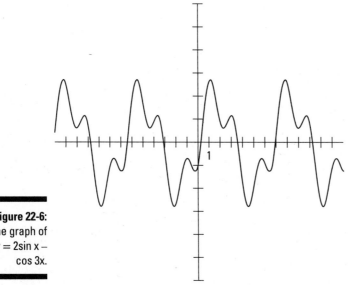

Figure 22-6:
The graph of
y = 2sin x –
cos 3x.

Applying Yourself to the Task

The graphs of some of the trig functions that you can create by altering the functions or combining them are fun to look at. They may even be useful when you're preparing a special border or other artwork. But the practical uses of these graphs are what you consider in this section. A cardiologist looks at a graph of the heart's function and detects whether it's beating properly. The graphs of earthquake activity are of special interest to those hoping to predict the next one — with enough time to warn everyone.

Measuring the tide

Along the coast, the tides are of particular interest. The tides are affected by the gravitational pull of both the moon and the sun. The high tides and low tides follow a periodic pattern that you can model with the sine function. On a particular winter day, the high tide in Boston, Massachusetts, occurred at midnight. To determine the height of the water in the harbor, use the equation $H(t) = 4.8 \sin\left[\frac{\pi}{6}(t+3)\right] + 5.1$, where t represents the number of hours since midnight.

1. **Input the value of t into the equation.**

   At midnight, the value of t is 0. Putting 0 in for t in the equation gives you

   $$H(0) = 4.8\sin\left[\frac{\pi}{6}(3)\right] + 5.1$$
   $$= 4.8\sin\frac{\pi}{2} + 5.1$$
   $$= 4.8(1) + 5.1 = 9.9$$

   The greatest value of sine occurs at $\frac{\pi}{2}$, so it makes sense that high tide would be when the formula uses the sine of that value.

2. **Determine the altitude.**

   The multiplier of 4.8 is the amplitude — how far above and below the middle value the graph goes. The tides go 4.8 feet above and below the average amount on this particular day. The number added on at the end, 5.1, is the average height for the tides. So, the tide goes up to 9.9 feet and down to 0.3 feet — wading depth.

3. Find the period of the function.

The multiplier of $\frac{\pi}{6}$ affects the period. The period of the sine function is usually 2π. Divide 2π by $\frac{\pi}{6}$, and you get $2\pi \div \frac{\pi}{6} = 2\pi \cdot \frac{6}{\pi} = 12$. The period is 12 hours, so you know that the tides go through their entire cycle in 12 hours. The 3 added to the *t* is a shift horizontally; that number determines what times of day the high tide and low tide occur. Figure 22-7 shows a graph of this function and the different stages of the tide at different times. By looking at the graph, you can plan your sailing and clam-digging activities.

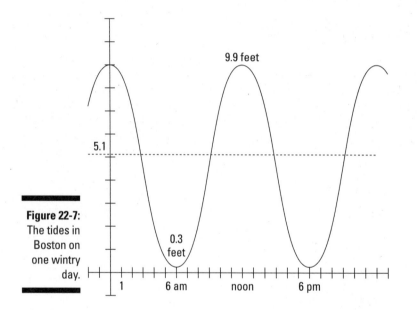

Figure 22-7: The tides in Boston on one wintry day.

Tracking the deer population

The graph in this section shows the population of a herd of deer, starting at the first of April and ending at the next April. New deer are born in the spring, so an increase in the herd size is expected. Predators take care of the weak deer — both young and old. And then you have to consider the weather; winter can be very hard on the population. Look at the graph in Figure 22-8 and see if it demonstrates what you'd expect. This cycle is the result of finding the sum of two different sine functions: $D(m) = 400 + 40 \sin(0.524m) + 20 \sin(1.047m)$ represents the population of the herd, where *m* is the number of months since April.

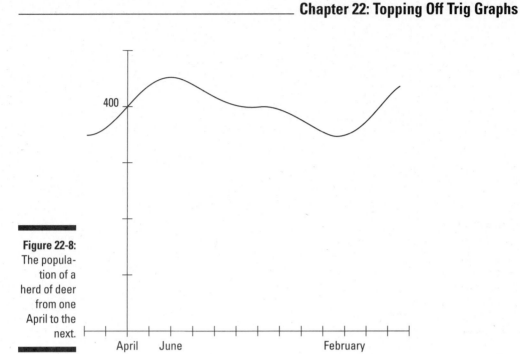

Figure 22-8:
The popula-
tion of a
herd of deer
from one
April to the
next.

400

April   June   February

The herd experiences a high of about 450 deer in June and a low of about 350 deer in February. This model shows a herd that stays pretty close to being the same size year after year.

Measuring the movement of an object on a spring

In this section, a trig function proves useful in a model for an object attached to a spring. The same pattern doesn't occur over and over, as it does in the previous sections. But this is a great example of a trig function at work.

The equation $H(t) = 3(0.7)^t \cos 5t + 4$ represents the height of an object attached to a spring, where t is the amount of time that has passed — usually in seconds. The equation has a trig function multiplied by an exponential function. When you first release the spring, the object hits a height of about 7 feet. It jumps up and down, finally settling in at about 4 feet high, as shown in Figure 22-9.

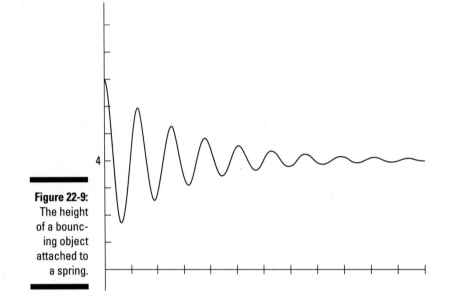

Figure 22-9:
The height
of a bounc-
ing object
attached to
a spring.

You can probably come up with a similar model to show how a bungee jumper goes up and down. So, you see, the trig functions have all sorts of applications — many of them very useful.

Part VI
The Part of Tens

For a list of ten ways to compute trig functions without trig functions, head to www.dummies.com/extras/trigonometry.

In this part...

- ✔ Create alternate versions of the basic trig identities.
- ✔ Look at some not-so-simple identities that can be useful in science applications.
- ✔ Take Pythagoras to another level with new versions of the identities.
- ✔ Change sums to products and products to sums.

Chapter 23

Ten Basic Identities . . . Plus Some Bonuses

In This Chapter

▶ Lining up the reciprocal, ratio, Pythagorean, and opposite-angle identities

▶ Tweaking the basic identities

▶ Using building blocks to manipulate trig expressions

A big advantage of trig expressions and equations is that you can adjust them in so many ways to suit your needs. The basic identities that I list in this chapter are the ones people use most frequently (and remember most often). And you'll also find some alternate notation and optional formats.

Reciprocal Identities

Take a look at the first reciprocal identity and its counterpart:

$$\sin\theta = \frac{1}{\csc\theta} \quad \text{and} \quad \csc\theta = \frac{1}{\sin\theta}$$

An alternate way of writing these identities uses an exponent of –1 rather than a fraction:

$$\sin\theta = (\csc\theta)^{-1} \quad \text{and} \quad \csc\theta = (\sin\theta)^{-1}$$

Note that the exponents apply to the entire function. These are not the inverse functions: $\csc^{-1}\theta$ and $\sin^{-1}\theta$.

Secant, cosecant, and cotangent are technically the three reciprocal functions, but you can write identities to show their reciprocals, too. Next are the second reciprocal identity and its counterpart.

$$\cos\theta = \frac{1}{\sec\theta} \quad \text{and} \quad \sec\theta = \frac{1}{\cos\theta}$$

Again, another way of writing these is to use an exponent of –1. The parentheses are used to be sure you recognize that this is the reciprocal, not the inverse.

$\cos\theta = (\sec\theta)^{-1}$ and $\sec\theta = (\cos\theta)^{-1}$

The tangent and its reciprocal at least have names that sound alike. The other two basic functions and their reciprocals (see the preceding equations) don't seem to have names that are not as nicely related.

$\tan\theta = \dfrac{1}{\cot\theta}$ and $\cot\theta = \dfrac{1}{\tan\theta}$

And, to finish off the alternate notation:

$\tan\theta = (\cot\theta)^{-1}$ and $\cot\theta = (\tan\theta)^{-1}$

Ratio Identities

Both of the ratio identities involve fractions with sine and cosine.

$$\tan\theta = \frac{\sin\theta}{\cos\theta}$$

$$\cot\theta = \frac{\cos\theta}{\sin\theta}$$

Here's a helpful way to remember which ratio identity has the sine in the numerator: Tangent and sine have beginning letters that are very close in the alphabet, and cotangent and cosine have the same beginning letters. This train of thought helped me out in high school when I first saw these identities.

Sometimes you want to have tangent or cotangent written in terms of either sine or cosine, not both. The following show one version of each. The denominators come from rewriting a Pythagorean identity. See the next section for more.

$$\tan\theta = \frac{\sin\theta}{\pm\sqrt{1-\sin^2\theta}}$$

$$\cot\theta = \frac{\cos\theta}{\pm\sqrt{1-\cos^2\theta}}.$$

Pythagorean Identities

The first Pythagorean identity uses your good friends sine and cosine and is probably one of the most frequently used identities.

$$\sin^2 \theta + \cos^2 \theta = 1$$

This second Pythagorean identity comes from the first Pythagorean identity; simply divide each term in that identity by the square of the cosine function and simplify.

$$\tan^2 \theta + 1 = \sec^2 \theta$$

Last but certainly not least, you get the third Pythagorean identity from the first one by dividing each of the terms in that identity by the square of the sine.

$$1 + \cot^2 \theta = \csc^2 \theta$$

You often come across instances where you want an expression equal to $\sin^2 \theta$ or $\cos^2 \theta$ or $\tan^2 \theta$ or $\cot^2 \theta$. And, when you do use the basic Pythagorean identities to solve for these expressions, you end up with a very handy and useful difference of squares. Here are some alternate versions of these basic identities:

$$\sin^2 \theta = 1 - \cos^2 \theta$$

$$\cos^2 \theta = 1 - \sin^2 \theta$$

$$\tan^2 \theta = \sec^2 \theta - 1$$

$$\cot^2 \theta = \csc^2 \theta - 1$$

Opposite-Angle Identities

This opposite-angle identity allows you to change to a positive angle when using the sine function:

$$\sin(-\theta) = -\sin \theta$$

This second opposite-angle identity may seem a little odd — the fact that the cosine of a negative angle has the same value as the cosine of the corresponding positive angle is always a little puzzling at first.

$$\cos(-\theta) = \cos \theta$$

The opposite-angle identity for the tangent follows the same pattern as that of the sine:

$\tan(-\theta) = -\tan \theta$.

Multiple-Angle Identities

And now for some more bonus identities. If you need an identity for the function sin $3x$ or cos $4x$, you can always create them yourself by applying an addition or double-angle formula. For example, sin $3x$ can be written as sin $(2x + x)$ and cos $4x$ can be cos $(2 \cdot 2x)$. Just to save you the trouble, here are a few of these special identities:

$\sin 3\theta = 3 \sin \theta - 4 \sin^3 \theta$

$\cos 3\theta = 4 \cos^3 \theta - 3 \cos \theta$

$\tan 3\theta = \dfrac{3 \tan \theta - \tan^3 \theta}{1 - 3 \tan^2 \theta}$

$\sin 4\theta = 8 \cos^3 \theta \sin \theta - 4 \cos \theta \sin \theta$

$\cos 4\theta = 8 \cos^4 \theta - 8 \cos^2 \theta + 1$

$\tan 4\theta = \dfrac{4 \tan \theta - 4 \tan^3 \theta}{1 - 6 \tan^2 \theta + \tan^4 \theta}$

Chapter 24

Ten Not-So-Basic Identities

*I*n Chapters 11 and 12, I cover the most frequently used identities at great length. Here are ten identities that don't appear in those chapters, because you won't use them all that often. A few are rather obscure. These identities don't lend themselves to memorization very well — you'll be better off just looking them up if you need them.

Product-to-Sum Identities

The product-to-sum identities look very much alike. You have to pay close attention to the subtle differences so that you can apply them correctly. Even though the product looks nice and compact, it's not always as easy to deal with in calculus computations — the sum or difference of two different angles is preferred.

The first identity has two angles, *A* and *B*. When you multiply the sine of one angle times the cosine of the other angle, you end up with one-half the sum of a sum identity and a difference identity. Whew!

$$\sin A \cos B = \frac{1}{2}\left[\sin(A+B) + \sin(A-B)\right]$$

This time, multiply the sines of both angles together, and the result equals one-half the difference between a sum identity and a difference identity:

$$\sin A \sin B = \frac{1}{2}\left[\cos(A-B) - \cos(A+B)\right]$$

This identity has a mix-and-match feel to it. Two different angles and two different functions are used. There seems to be something for everyone.

$$\cos A \sin B = \frac{1}{2}\left[\sin(A+B)-\sin(A-B)\right]$$

The last product-to-sum identity uses the cosines of two angles:

$$\cos A \cos B = \frac{1}{2}\left[\cos(A-B)+\cos(A+B)\right]$$

Just in case you think this is hocus-pocus or that I'm making these up, let me show you an example of one of these new identities. Using $A = 45$ degrees and $B = 30$ degrees and the identity $\cos A \sin B = \frac{1}{2}\left[\sin(A+B)-\sin(A-B)\right]$,

$$\cos 45° \sin 30° = \frac{1}{2}\left[\sin(45°+30°)-\sin(45°-30°)\right]$$

$$\cos 45° \sin 30° = \frac{1}{2}[\sin(75°)-\sin(15°)]$$

$$\frac{\sqrt{2}}{2}\cdot\frac{1}{2} = \frac{1}{2}\left[\frac{\sqrt{6}+\sqrt{2}}{4}-\frac{\sqrt{6}-\sqrt{2}}{4}\right]$$

Where did I get those values for the sine of 75 and 15 degrees? I found the sine of 75 degrees back in Chapter 12. For 15 degrees, I used the sine of the difference between 45 degrees and 30 degrees. Now, simplifying,

$$\frac{\sqrt{2}}{4} = \frac{1}{2}\left[\frac{\cancel{\sqrt{6}}+\sqrt{2}-\cancel{\sqrt{6}}+\sqrt{2}}{4}\right]$$

$$= \frac{1}{\cancel{2}}\left[\frac{\cancel{2}\sqrt{2}}{4}\right] = \frac{\sqrt{2}}{4}$$

Sum-to-Product Identities

The sum to product identities are useful for modeling what happens with sound frequencies. Think of two different tones represented by sine curves. Add them together, and they beat against each other with a warble — how much depends on their individual frequencies. The identities give a function modeling what's happening.

The first identity takes two different angles, A and B, and adds their sines together. The result: twice the product of the sine and cosine of two new angles that are created by halving the sum and difference of the angles. See for yourself:

$$\sin A + \sin B = 2\sin\left(\frac{A+B}{2}\right)\cos\left(\frac{A-B}{2}\right)$$

You can technically call this next identity a difference-to-product identity, although math gurus usually classify it with the sum-to-product identities. Of course, you can consider the difference to be a sum if you call it the sum of a sine and the opposite of another sine.

$$\sin A - \sin B = 2\cos\left(\frac{A+B}{2}\right)\sin\left(\frac{A-B}{2}\right)$$

This next identity involves the sum of the cosines of two angles.

$$\cos A + \cos B = 2\cos\left(\frac{A+B}{2}\right)\cos\left(\frac{A-B}{2}\right)$$

As you probably expect, the last sum-to-product identity has the difference of the cosines of two angles.

$$\cos A - \cos B = -2\sin\left(\frac{A+B}{2}\right)\sin\left(\frac{A-B}{2}\right)$$

For a look at how you use these identities, I show you the difference of the cosines of angles $A = 60$ and $B = 30$.

$$\cos A - \cos B = -2\sin\left(\frac{A+B}{2}\right)\sin\left(\frac{A-B}{2}\right)$$

$$\cos 60° - \cos 30° = -2\sin\left(\frac{60° + 30°}{2}\right)\sin\left(\frac{60° - 30°}{2}\right)$$

$$\cos 60° - \cos 30° = -2\sin\left(45°\right)\sin\left(15°\right)$$

$$\frac{1}{2} - \frac{\sqrt{3}}{2} = -2\left(\frac{\sqrt{2}}{2}\right)\left(\frac{\sqrt{6} - \sqrt{2}}{4}\right)$$

I picked up the sine of 15 degrees from the previous section. Simplifying,

$$\frac{1-\sqrt{3}}{2} = -\cancel{2}\left(\frac{\sqrt{2}}{\cancel{2}}\right)\left(\frac{\sqrt{6} - \sqrt{2}}{4}\right)$$

$$= -\frac{\sqrt{2}\left(\sqrt{6} - \sqrt{2}\right)}{4} = -\frac{\sqrt{12} - \sqrt{4}}{4}$$

$$= -\frac{2\sqrt{3} - 2}{4} = -\frac{\cancel{2}\left(\sqrt{3} - 1\right)}{\cancel{4}^2}$$

$$= \frac{1 - \sqrt{3}}{2}$$

Reduction Formula

The reduction formula reduces two trig functions into one. It's useful when studying the force of a spring, the position of a swinging pendulum, or the current in an electrical circuit. This formula takes the sum of two different functions with the common input x and changes the sum to a single function where the multiplier of the sine is the amplitude and the phase shift is θ. The a and b are the coordinates of some point on the terminal side of θ when it's in standard position:

$$a\sin x + b\cos x = \sqrt{a^2+b^2}\,\sin(x+\theta)$$

To get the sine and cosine of x, you can use simplified versions:

$$\sin x = \frac{b}{\sqrt{a^2+b^2}}\text{ and }\cos x = \frac{a}{\sqrt{a^2+b^2}}$$

Mollweide's Equations

Karl Mollweide was an astronomy teacher. His work in astronomy and mathematics led to his discovery of identities that can take the measures of the sides of a triangle and relate them to an expression involving trig functions. Mollweide's equations involve all six parts of a triangle: the three angles, A, B, and C, and the three corresponding sides opposite those angles, a, b, and c.

$$\frac{a+b}{c}=\frac{\cos\left(\dfrac{A-B}{2}\right)}{\sin\left(\dfrac{C}{2}\right)}\text{ or }\frac{a-b}{c}=\frac{\sin\left(\dfrac{A-B}{2}\right)}{\cos\left(\dfrac{C}{2}\right)}$$

Showing an example of one of these proportions, I use a 30-60-90 right triangle, with sides 1, $\sqrt{3}$ and 2, because the numbers are so nice. The side opposite angle $A = 30$ degrees measures 1. The side opposite angle $B = 60$ degrees measures $\sqrt{3}$. And the side opposite the right angle measures 2. Using the left-hand equation,

$$\frac{a+b}{c}=\frac{\cos\left(\dfrac{A-B}{2}\right)}{\sin\left(\dfrac{C}{2}\right)}=\frac{\cos\left(\dfrac{30°-60°}{2}\right)}{\sin\left(\dfrac{90°}{2}\right)}$$

$$=\frac{\cos(-15°)}{\sin(45°)}=\frac{\cos(15°)}{\sin(45°)}$$

$$\frac{1+\sqrt{3}}{2}=\frac{\dfrac{\sqrt{6}+\sqrt{2}}{4}}{\dfrac{\sqrt{2}}{2}}=\frac{\sqrt{6}+\sqrt{2}}{\cancel{4}^{2}}\cdot\frac{\cancel{2}}{\sqrt{2}}$$

$$=\frac{\sqrt{6}+\sqrt{2}}{2\sqrt{2}}\cdot\frac{\sqrt{2}}{\sqrt{2}}=\frac{\sqrt{12}+\sqrt{4}}{2\sqrt{4}}$$

$$=\frac{2\sqrt{3}+2}{4}=\frac{\cancel{2}\left(\sqrt{3}+1\right)}{\cancel{4}^{2}}=\frac{1+\sqrt{3}}{2}$$

Appendix
Trig Functions Table

θ	$\sin\theta$	$\cos\theta$	$\tan\theta$	$\cot\theta$	$\sec\theta$	$\csc\theta$
0°	0.000	1.000	0.000	Undefined	1.000	Undefined
1°	0.017	1.000	0.017	57.290	1.000	57.299
2°	0.035	0.999	0.035	28.636	1.001	28.654
3°	0.052	0.999	0.052	19.081	1.001	19.107
4°	0.070	0.998	0.070	14.301	1.002	14.336
5°	0.087	0.996	0.087	11.430	1.004	11.474
6°	0.105	0.995	0.105	9.514	1.006	9.567
7°	0.122	0.993	0.123	8.144	1.008	8.206
8°	0.139	0.990	0.141	7.115	1.010	7.185
9°	0.156	0.988	0.158	6.314	1.012	6.392
10°	0.174	0.985	0.176	5.671	1.015	5.759
11°	0.191	0.982	0.194	5.145	1.019	5.241
12°	0.208	0.978	0.213	4.705	1.022	4.810
13°	0.225	0.974	0.231	4.331	1.026	4.445
14°	0.242	0.970	0.249	4.011	1.031	4.134
15°	0.259	0.966	0.268	3.732	1.035	3.864
16°	0.276	0.961	0.287	3.487	1.040	3.628
17°	0.292	0.956	0.306	3.271	1.046	3.420
18°	0.309	0.951	0.325	3.078	1.051	3.236
19°	0.326	0.946	0.344	2.904	1.058	3.072
20°	0.342	0.940	0.364	2.747	1.064	2.924
21°	0.358	0.934	0.384	2.605	1.071	2.790
22°	0.375	0.927	0.404	2.475	1.079	2.669
23°	0.391	0.921	0.424	2.356	1.086	2.559
24°	0.407	0.914	0.445	2.246	1.095	2.459

(continued)

θ	$\sin\theta$	$\cos\theta$	$\tan\theta$	$\cot\theta$	$\sec\theta$	$\csc\theta$
25°	0.423	0.906	0.466	2.145	1.103	2.366
26°	0.438	0.899	0.488	2.050	1.113	2.281
27°	0.454	0.891	0.510	1.963	1.122	2.203
28°	0.469	0.883	0.532	1.881	1.133	2.130
29°	0.485	0.875	0.554	1.804	1.143	2.063
30°	0.500	0.866	0.577	1.732	1.155	2.000
31°	0.515	0.857	0.601	1.664	1.167	1.972
32°	0.530	0.848	0.625	1.600	1.179	1.887
33°	0.545	0.839	0.649	1.540	1.192	1.836
34°	0.559	0.829	0.675	1.483	1.206	1.788
35°	0.574	0.819	0.700	1.428	1.221	1.743
36°	0.588	0.809	0.727	1.376	1.236	1.701
37°	0.602	0.799	0.754	1.327	1.252	1.662
38°	0.616	0.788	0.781	1.280	1.269	1.624
39°	0.629	0.777	0.810	1.235	1.287	1.589
40°	0.643	0.766	0.839	1.192	1.305	1.556
41°	0.656	0.755	0.869	1.150	1.325	1.524
42°	0.669	0.743	0.900	1.111	1.346	1.494
43°	0.682	0.731	0.933	1.072	1.367	1.466
44°	0.695	0.719	0.966	1.036	1.390	1.440
45°	0.707	0.707	1.000	1.000	1.414	1.414
46°	0.719	0.695	1.036	0.966	1.440	1.390
47°	0.731	0.682	1.072	0.933	1.466	1.367
48°	0.743	0.669	1.111	0.900	1.494	1.346
49°	0.755	0.656	1.150	0.869	1.524	1.325
50°	0.766	0.643	1.192	0.839	1.556	1.305
51°	0.777	0.629	1.235	0.810	1.589	1.287
52°	0.788	0.616	1.280	0.781	1.624	1.269
53°	0.799	0.602	1.327	0.754	1.662	1.252
54°	0.809	0.588	1.376	0.727	1.701	1.236
55°	0.819	0.574	1.428	0.700	1.743	1.221
56°	0.829	0.559	1.483	0.675	1.788	1.206
57°	0.839	0.545	1.540	0.649	1.836	1.192

θ	$\sin\theta$	$\cos\theta$	$\tan\theta$	$\cot\theta$	$\sec\theta$	$\csc\theta$
58°	0.848	0.530	1.600	0.625	1.887	1.179
59°	0.857	0.515	1.664	0.601	1.972	1.167
60°	0.866	0.500	1.732	0.577	2.000	1.155
61°	0.875	0.485	1.804	0.554	2.063	1.143
62°	0.883	0.469	1.881	0.532	2.130	1.133
63°	0.891	0.454	1.963	0.510	2.203	1.122
64°	0.899	0.438	2.050	0.488	2.281	1.113
65°	0.906	0.423	2.145	0.466	2.366	1.103
66°	0.914	0.407	2.246	0.445	2.459	1.095
67°	0.921	0.391	2.356	0.424	2.559	1.086
68°	0.927	0.375	2.475	0.404	2.669	1.079
69°	0.934	0.358	2.605	0.384	2.790	1.071
70°	0.940	0.342	2.747	0.364	2.924	1.064
71°	0.946	0.326	2.904	0.344	3.072	1.058
72°	0.951	0.309	3.078	0.325	3.236	1.051
73°	0.956	0.292	3.271	0.306	3.420	1.046
74°	0.961	0.276	3.487	0.287	3.628	1.040
75°	0.966	0.259	3.732	0.268	3.864	1.035
76°	0.970	0.242	4.011	0.249	4.134	1.031
77°	0.974	0.225	4.331	0.231	4.445	1.026
78°	0.978	0.208	4.705	0.213	4.810	1.022
79°	0.982	0.191	5.145	0.194	5.241	1.019
80°	0.985	0.174	5.671	0.176	5.759	1.015
81°	0.988	0.156	6.314	0.158	6.392	1.012
82°	0.990	0.139	7.115	0.141	7.185	1.010
83°	0.993	0.122	8.144	0.123	8.206	1.008
84°	0.995	0.105	9.514	0.105	9.567	1.006
85°	0.996	0.087	11.430	0.087	11.474	1.004
86°	0.998	0.070	14.301	0.070	14.336	1.002
87°	0.999	0.052	19.081	0.052	19.107	1.001
88°	0.999	0.035	28.636	0.035	28.654	1.001
89°	1.000	0.017	57.290	0.017	57.299	1.000
90°	1.000	0.000	Undefined	0.000	Undefined	1.000

Index

About the Author

Mary Jane Sterling is also the author of *Algebra I For Dummies, Algebra II For Dummies, Math Word Problems For Dummies, Business Math For Dummies*, and *Linear Algebra For Dummies*. She taught junior high school and high school math for several years before beginning her current 30-plus-year tenure at Bradley University in Peoria, Illinois. Mary Jane especially enjoys working with future teachers and trying out new technology. She and her husband, Ted, are enjoying their time with children, grandchildren, and fishing (not necessarily in that order).

Dedication

I would like to dedicate *Trigonometry For Dummies* to the Kiwanis Club of Peoria (Downtown). Working with this club over the past 20 years has allowed me to have wonderful interactions with members and participate in projects with the area Aktion Clubs (for special-needs adults), Bradley Circle K (college service organization), and Key Clubs (high school groups). The friendships and fellowship have enriched my life immensely, just as the club's service has enriched the community.

Author's Acknowledgments

I give a huge thank-you to my project editor, Elizabeth Kuball, who has agreed, yet again, to take on a challenging mathematics book. Elizabeth is always so efficient yet understanding, professional yet forgiving, and a peach to work with under pressure.

Thank you to my technical editor, Shira Fass. As careful as I try to be, I still seem to miss one thing or another. Thank you for catching any silly (or not so silly) errors.

And, of course, a grateful thank you to my acquisitions editor, Lindsay Lefevere, who seems to like to keep me busy — which is a very good thing.

Publisher's Acknowledgments

Executive Editor: Lindsay Sandman Lefevere

Project Editor: Elizabeth Kuball

Copy Editor: Elizabeth Kuball

Technical Editor: Shira Fass

Project Coordinator: Erin Zeltner

Cover Image: © Peter Hermes Furian/Alamy

Apple & Mac

iPad For Dummies,
5th Edition
978-1-118-49823-1

iPhone 5 For Dummies,
6th Edition
978-1-118-35201-4

MacBook For Dummies,
4th Edition
978-1-118-20920-2

OS X Mountain Lion
For Dummies
978-1-118-39418-2

Blogging & Social Media

Facebook For Dummies,
4th Edition
978-1-118-09562-1

Mom Blogging
For Dummies
978-1-118-03843-7

Pinterest For Dummies
978-1-118-32800-2

WordPress For Dummies,
5th Edition
978-1-118-38318-6

Business

Commodities For Dummies,
2nd Edition
978-1-118-01687-9

Investing For Dummies,
6th Edition
978-0-470-90545-6

Personal Finance

Personal Finance
For Dummies,
7th Edition
978-1-118-11785-9

QuickBooks 2013
For Dummies
978-1-118-35641-8

Small Business Marketing Kit
For Dummies,
3rd Edition
978-1-118-31183-7

Careers

Job Interviews
For Dummies,
4th Edition
978-1-118-11290-8

Job Searching with
Social Media
For Dummies
978-0-470-93072-4

Personal Branding
For Dummies
978-1-118-11792-7

Resumes For Dummies,
6th Edition
978-0-470-87361-8

Success as a Mediator
For Dummies
978-1-118-07862-4

Diet & Nutrition

Belly Fat Diet For Dummies
978-1-118-34585-6

Eating Clean For Dummies
978-1-118-00013-7

Nutrition For Dummies,
5th Edition
978-0-470-93231-5

Digital Photography

Digital Photography
For Dummies,
7th Edition
978-1-118-09203-3

Digital SLR Cameras &
Photography For Dummies,
4th Edition
978-1-118-14489-3

Photoshop Elements 11
For Dummies
978-1-118-40821-6

Gardening

Herb Gardening
For Dummies,
2nd Edition
978-0-470-61778-6

Vegetable Gardening
For Dummies,
2nd Edition
978-0-470-49870-5

Health

Anti-Inflammation Diet
For Dummies
978-1-118-02381-5

Diabetes For Dummies,
3rd Edition
978-0-470-27086-8

Living Paleo For Dummies
978-1-118-29405-5

Hobbies

Beekeeping
For Dummies
978-0-470-43065-1

eBay For Dummies,
7th Edition
978-1-118-09806-6

Raising Chickens
For Dummies
978-0-470-46544-8

Wine For Dummies,
5th Edition
978-1-118-28872-6

Writing Young Adult Fiction
For Dummies
978-0-470-94954-2

Language &
Foreign Language

500 Spanish Verbs
For Dummies
978-1-118-02382-2

English Grammar
For Dummies,
2nd Edition
978-0-470-54664-2

French All-in One
For Dummies
978-1-118-22815-9

German Essentials
For Dummies
978-1-118-18422-6

Italian For Dummies
2nd Edition
978-1-118-00465-4

ⓔ **Available in print and e-book formats.**

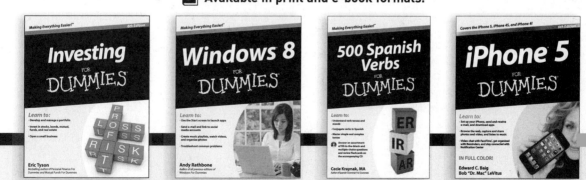

Math & Science

Algebra I For Dummies,
2nd Edition
978-0-470-55964-2

Anatomy and Physiology
For Dummies,
2nd Edition
978-0-470-92326-9

Astronomy For Dummies,
3rd Edition
978-1-118-37697-3

Biology For Dummies,
2nd Edition
978-0-470-59875-7

Chemistry For Dummies,
2nd Edition
978-1-1180-0730-3

Pre-Algebra Essentials
For Dummies
978-0-470-61838-7

Microsoft Office

Excel 2013 For Dummies
978-1-118-51012-4

Office 2013 All-in-One
For Dummies
978-1-118-51636-2

PowerPoint 2013
For Dummies
978-1-118-50253-2

Word 2013 For Dummies
978-1-118-49123-2

Music

Blues Harmonica
For Dummies
978-1-118-25269-7

Guitar For Dummies,
3rd Edition
978-1-118-11554-1

iPod & iTunes
For Dummies,
10th Edition
978-1-118-50864-0

Programming

Android Application
Development For
Dummies, 2nd Edition
978-1-118-38710-8

iOS 6 Application
Development For Dummies
978-1-118-50880-0

Java For Dummies,
5th Edition
978-0-470-37173-2

Religion & Inspiration

The Bible For Dummies
978-0-7645-5296-0

Buddhism For Dummies,
2nd Edition
978-1-118-02379-2

Catholicism For Dummies,
2nd Edition
978-1-118-07778-8

Self-Help & Relationships

Bipolar Disorder
For Dummies,
2nd Edition
978-1-118-33882-7

Meditation For Dummies,
3rd Edition
978-1-118-29144-3

Seniors

Computers For Seniors
For Dummies,
3rd Edition
978-1-118-11553-4

iPad For Seniors
For Dummies,
5th Edition
978-1-118-49708-1

Social Security
For Dummies
978-1-118-20573-0

Smartphones & Tablets

Android Phones
For Dummies
978-1-118-16952-0

Kindle Fire HD
For Dummies
978-1-118-42223-6

NOOK HD For Dummies,
Portable Edition
978-1-118-39498-4

Surface For Dummies
978-1-118-49634-3

Test Prep

ACT For Dummies,
5th Edition
978-1-118-01259-8

ASVAB For Dummies,
3rd Edition
978-0-470-63760-9

GRE For Dummies,
7th Edition
978-0-470-88921-3

Officer Candidate Tests,
For Dummies
978-0-470-59876-4

Physician's Assistant Exam
For Dummies
978-1-118-11556-5

Series 7 Exam
For Dummies
978-0-470-09932-2

Windows 8

Windows 8 For Dummies
978-1-118-13461-0

Windows 8 For Dummies,
Book + DVD Bundle
978-1-118-27167-4

Windows 8 All-in-One
For Dummies
978-1-118-11920-4

e **Available in print and e-book formats.**

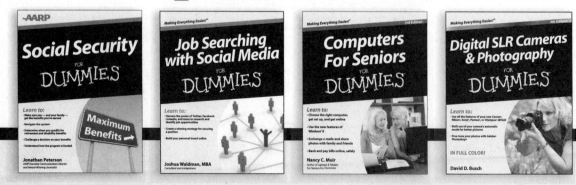